HISTORY OF THE UNITED STATES

MARINE CORPS

VMA-231
MARINES
CG
158390
MARINES
VMA-231

HISTORY OF THE UNITED STATES

MARINE CORPS

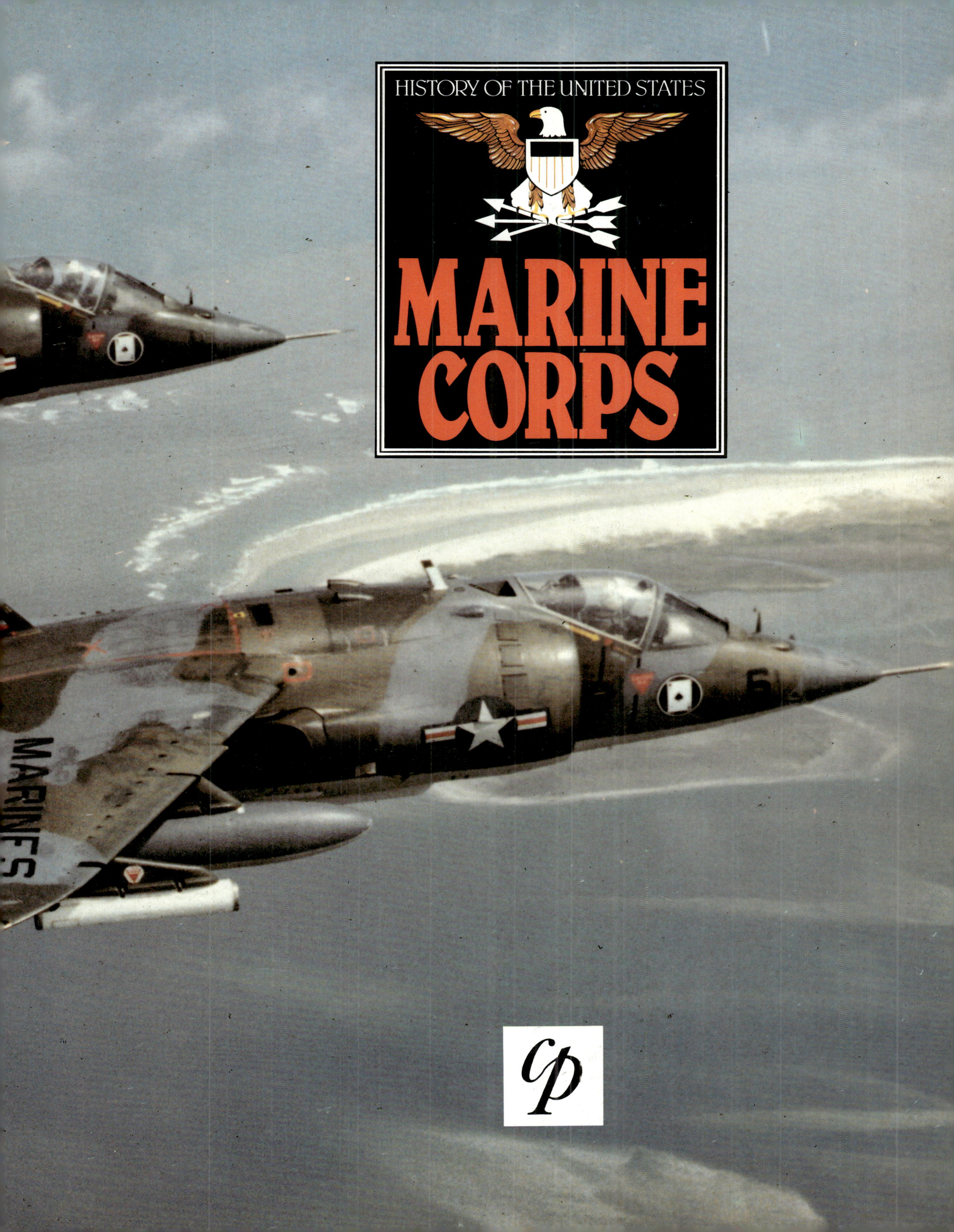

First published by
Chevprime Limited
27 Swinton Street
London WC1

Printed in Italy

ISBN 1-85361-063-1

Right: Jungle warfare is one of the Marines' special skills.

1:FIRST TO FIGHT 1775-1914

THE AMERICAN WAR OF INDEPENDENCE

The origins of the Marine Corps can be traced back to the colonial battalions of Marines raised in America before the Revolution and modeled on their British counterparts.

In 1740 four battalions of Marines — some 3,000 men — were raised for service in Britain's war with Spain, the so-called "War of Jenkins' Ear", allegedly triggered off by the arrest and mutilation of a British captain. Designated 43rd Regiment of Foot, they became known as "Gooch's Marines", after their commander, Colonel William Gooch. Serving under the British Admiral Vernon, Colonel Gooch led his Marines into battle at Cartagena, Colombia, in April 1741. In July of the same year, Marines landed unopposed at Guantanamo Bay, in Cuba, to secure it as a base for a British fleet.

The War of Jenkins' Ear soon expanded into a Franco-British confrontation coinciding with, and merging into, the War of the Spanish Succession. In this new struggle, the Colonial Marines found employment in the British Navy, serving as marksmen aboard its ships and providing the cutting edge of boarding parties. Their duties also included shipboard security and maintaining discipline among the seamen, the latter a task calculated to encourage the sometimes prickly relationship between Marines and sailors, a phenomenon which persists today.

The Continental Marines

The British soldier has lost many battles, but the only important war he has lost is the American War of Independence, which began with shots on the village green at Lexington on 19 April 1775. The unity of opposition felt by the American colonies to British rule had already been demonstrated at the First Continental Congress, which met in Philadelphia in September 1774. In May 1775, the Second Continental Congress prepared for war, appointing George Washington on 14 June as commander-in-chief of the army to defend the 13 colonies.

Initially, this predominantly militia force did not fare very well against British regulars. An invasion of Canada (June 1775–July 1776) proved beyond their resources and

Left: George Washington at Valley Forge.

came to grief at the end of December 1775 when an attempt to storm Quebec was repulsed and the colonists' commander, General Montgomery, was killed.

Two months earlier, the Congress had approved a resolution calling for the creation of a new military force, the Continental Marines. The resolution had been drafted by a committee of Congress in the Tun Tavern, Philadelphia, whose owner, Robert Mullan, was appointed a Marine Captain under the command of another inn-keeper, Samuel Nicholas, proprietor of The Conestoga Wagon.

However, Marines had already made their mark on the war. In May 1775, the Continental Congress reacted to news of the enfeebled state of the American forces holding Fort Ticonderoga and Crown Point by requesting the governor of Connecticut to reinforce the garrisons with his colony's forces. A body of troops set out from Hartford, and with them was "money escorted by eight Marines [from the Connecticut Navy] well spirited and equipped". The Marines — later dubbed "The Original Eight" — and the soldiers evaded hostile Indians and British forces to relieve the hard-pressed garrison.

In October 1775, American merchantmen were requisitioned and armed in a successful bid to capture two unescorted British ships carrying quantities of ammunition, of which Washington's forces were in acutely short supply. The vessels were placed on "Continental risque and pay", an arrangement whereby Congress paid the crews and assumed responsibility for insuring the ships against loss or capture. Congress also instructed Washington to "give proper encouragement to the Marines and seamen" — the first written reference by Congress to "Marines".

The Continental Marines went into action in March 1776 when an American squadron raided the Bahamas with the principal aim of seizing gunpowder for Washington's army. On 3 March, 220 Marines, under the command of Captain Samuel Nicholas, landed on New Providence Island. In two weeks the raiding party captured two forts, a quantity of guns, occupied Nassau and took the British Governor prisoner. But the gunpowder had been spirited away by the canny Governor Browne. Nicholas reported, "We found in this fort [Nassau] a great quantity of shot and shells, with 15 brass mortars, but the grand article, gunpowder, the Governor had sent off before, viz 150 casks." The Marines re-embarked on 17 March for Rhode Island. On the homeward journey the squadron was engaged by the lone 20-gun British corvette *Glasgow*. In a night action with the American *Cabot* and the squadron's flagship, the 24-gun *Alfred*, the British inflicted the first combat losses suffered by the Marines — Second Lieutenant John Fitzpatrick and six enlisted men.

The expedition earned Nicholas promotion to the rank of Major, and in December 1776 he and 300 of his men joined Washington's army in Pennsylvania. The Marines did not participate in Washington's morale-boosting victory at Trenton on 26 January, remaining on the Pennsylvania side of the River Delaware as a reserve force. Nevertheless, this was the first occasion on which Marines served as part of the Army.

Right: The capture of the British Ship *Sandwich*, during the "undeclared" naval war with France, by Captain Daniel Carmick of the Marine detachment of the USS *Constitution*, 11 May 1800.

The Marines fought in the second battle of Trenton (2 January 1777), using the ruse of false campfires to slip through General Charles Cornwallis' lines, and took part in the action at Princeton on the following day.

In the spring of 1777, Washington reorganized his army, incorporating some of the Marines in his artillery and returning the remainder to their naval duties. In October–November 1777, the Marine artillerymen played a prominent part in the defense of Fort Mifflin, on the River Delaware, against a squadron of British warships and Hessian shore batteries.

In January 1778, a small force of Marines, commanded by Captain James Willing, sailed down the Mississippi aboard a venerable vessel which Willing had renamed *Rattletrap*. During the next 12 months Willing's force operated in the region of New Orleans, harassing British traders, before returning north to fight against hostile Indians.

In July 1798, a force of 300 Marines, under Captain John Welsh, distinguished themselves in an unsuccessful combined operation against a British fort at Penobscot Bay, Maine, making "a forcible charge against the enemy". In May 1780, 200 Marines fought a gallant holding action against a vastly superior British force at Charleston, South Carolina.

Continental Marines serving with John Paul Jones aboard *Ranger* took the war to the enemy, making two raids on the British mainland in April 1778. In the winter of 1780–81, Marines from the frigate *South Carolina* made an amphibious attack on the island of Jersey in the English Channel, the last such operation of the war. On 8 May 1782, 300 Marines on the *South Carolina* assisted in the capture of the Bahama Islands for Spain. In January 1783 the Continental Marines ended their part in the naval war when a detachment aboard the warship *Hague* boarded and seized the British ship *Baille* in the West Indies.

The military outcome of the War of Independence had been effectively settled two years earlier, when General Burgoyne surrendered at Yorktown on 19 October 1781. Britain finally acknowledged the loss of her colonies at the Treaty of Versailles on 30 November 1783. With peace came the disbandment of the American Navy and the Continental Marines, whose peak strength during the war had been 124 officers and 3,000 men. Major Samuel Nicholas is held by the Marines to have been their first Commandant.

Birth of the Marine Corps

When the first Congress assembled in 1789 in New York, the United States had no Navy. A number of states maintained small warships, but the Republic had not a single armed vessel. It was not until March 1794 that Congress took steps to reactivate the Navy, authorizing the construction of six frigates, each of which was to have a detachment of Marines.

However, two years passed before the funds were found for the construction of three ships, *United States*, *Constellation* and *Constitution*, all three of which were launched in 1797. A Congressional Act of 1 July 1797 laid down the number of Marines to serve on the ships: five lieutenants, eight sergeants, eight corporals, three drummers, three fifers and 140 privates. At this point the Marines were considered to be part of the Navy.

An early supporter of the idea of a Marine Corps with its own separate identity and command structure was the Secretary of War, James McHenry, who in a letter of 9 April 1798 to the House Naval Committee Chairman, Samuel Sewell, recommended that an organization of Marines be formed. In Congress, wheels were grinding. The House of Representatives passed a Bill calling for the creation of "a battalion to be called the Marine Corps". The Senate increased the size of the proposed Corps to a regiment and on 11 July 1798, with House agreement, sent "An Act for Establishing and Organizing a Marine Corps" to President John Adams for his signature.

The Corps was initially placed under the direct orders of the President, to be attached either to the Army or the Navy, "according to the nature of the service in which they shall be employed" and subject either to the Articles of War or the Naval Regulations. The ambiguity inherent in this agreement — particularly when the Marines were serving ashore — was removed in June 1834 with the passing of an Act "For the Better Organization of the Marine Corps", which

Right: The engagement between the American ship *Planter* and a French privateer during the "undeclared" war with France, 10 July 1799.

attached the Marine Corps to the Navy unless the President ordered part of it to come temporarily under the control of the Army.

The table of organization for the new body of Marines prescribed a total of 33 officers and 848 "non-commissioned officers, musicians and privates". They were provided with striking uniforms: blue shortcoats and trousers, edged in brilliant red; their hats, with one side turned up, sported a yellow band and a cockade. Sergeants bore yellow epaulettes on their shoulders; officers wore long blue coats with red cuffs and golden epaulettes. All the uniforms had stiff leather collars, which quickly earned the Marines their famous nickname, the "Leathernecks".

On 12 July 1798, President Adams appointed William Ward Burrows as the first commandant of the Marine Corps. Burrows set up his first headquarters in Philadelphia, but in July 1800 moved the Marines HQ to Washington DC, selecting a location "near the Navy Yard and within easy marching distance of the Capitol". By 1806, Marine Barracks, Washington, was completed, and it is here that the Commandant of the Corps makes his home to this day.

War at Sea

One of the most pressing problems facing President Adams was the protection of American commerce against the depradations of French privateers, who were harassing the shipping trade between the United States and Britain, with whom France was at war. Persistent French disregard of American neutrality led to the "undeclared" naval war with France which began in May 1798.

Marines serving aboard Federal vessels played a prominent part in all the sea engagements of the war. They also participated in land action on foreign soil and were given the responsibility of guarding French prisoners of war held in Philadelphia.

In 1799, Marines serving on *Constellation*, under the command of Lieutenant Bartholomew Clinch, were involved in the capture of the French frigate *Insurgente*. In 1800 *Constellation* bested the French ship *Vengeance* in a four-hour night engagement. Between October 1799 and March 1800, the Marines aboard Navy ships saw action against the armed barges of Haitian pirates ("picaroons").

On 11 May 1800, sailors from *Constitution*, and its Marine detachment, seized the captured British ship *Sandwich*, held by the French in the supposedly neutral port of Puerto Plata on the northern coast of Santo Domingo. Employing a *ruse de guerre* reminiscent of the wooden horse of Troy, a party of sailors and 90 Marines sailed into the harbor hidden below-decks on the American ship *Sally*, took over *Sandwich*, captured the local fort, spiked its cannon and then made good their escape with their prize, the ship *Sandwich*. This was the first combat landing made by the Marine Corps on foreign soil. In September, Marines from *Patapsco* and *Merrimack* went into action against French units besieging the Dutch garrison at Willemstad on the island of Curacao. The arrival of the Marines forced the French to lift the siege.

One of the most successful of the US

Navy's ships was the *Enterprise*. In 1800 she defeated or captured 10 enemy vessels and recaptured 11 American ships. Her 16 Marines, delivering a hail of fire from the fighting tops, played an important part in all these victories.

The "undeclared" naval war came to an end in February 1801, by which time the US Navy had captured 85 French vessels. In 1802, however, President Thomas Jefferson instituted a policy of economic retrenchment aimed at reducing the national debt. He ordered the selling of naval vessels and halted all warship construction. The majority of the frigates which remained were dismantled to save expense. The Marines did not escape the financial axe, and the Corps' strength was reduced to 26 officers and 453 men.

Barbary Pirates

These economies were shortsighted measures at a time when the United States was about to enter into a war in the Mediterranean against the Barbary states. For many years the Muslim states of Morocco, Tunis, Algiers and Tripoli had been extorting money from the United States and the European powers, either as ransom for prisoners or as the price of allowing merchant ships to navigate the Mediterranean unmolested by the Barbary pirates. By the spring of 1801, the United States had parted with two million dollars, one-fifth of its national revenue.

Matters came to a head when the Pasha of Tripoli, Yusuf Caramanli, demanded even greater tribute. When this was refused, he declared war on the United States in May 1801. A month later, four vessels of the United States' exiguous Navy — *President*, *Philadelphia*, *Essex* and *Enterprise* — were formed into the Mediterranean Squadron and despatched from Hampton Roads to Tripoli. There were about 180 Marines in the Squadron.

The Mediterranean Squadron was far too small to protect American merchant ships or threaten the supremacy of the Barbary pirates. On 31 October 1803 the frigate *Philadelphia* ran aground on a reef off Tripoli, was captured by pirate ships, floated free and towed into port, where its crew (including 44 Marines) were imprisoned by the Pasha.

Philadelphia was saved from the ignominy of becoming a pirate ship by the daring of Lieutenant Stephen Decatur, USN. On the night of 16 February 1804, Decatur slipped into the harbor aboard the captured ketch *Intrepid*. With his raiding party were eight Marines under the command of Sergeant Solomon Wren. Decatur's party boarded *Philadelphia*, overpowered the pirate crew, burned the ship to the waterline, and in the subsequent confusion in the harbor escaped without suffering a single casualty. However, *Philadephia*'s crew remained in Yusuf's hands, and their freedom was purchased with another massive ransom.

One of the most remarkable episodes in the

Above left. Lieutenant Stephen Decatur, USN, who led a daring action against the Barbary Pirates in February 1804.
Above: Decatur and his raiding party board the captured frigate *Philadelphia* in Tripoli harbor prior to burning her.

Above: Lieutenant O'Bannon and his men raise the Stars and Stripes over Yusuf Caramanli's citadel at Derna in April 1805.

Tripoli war was the expedition undertaken by William Eaton, American Consul in Tunis, to replace the troublesome Yusuf with his elder brother, Hamet Bey. In Alexandria, Eaton assembled a force comprising a small number of US Navy and Marine Corps personnel, 40 Greeks, a squadron of Arab cavalry, 100 Turks and other mercenaries and a caravan of camels. In February 1805, he set out across the Libyan desert on a 600-mile trek to Derna, Yusuf's capital. Accompanying him was Marine Lieutenant Presley N. O'Bannon with a sergeant and six privates from the *Argus*.

During their seven-week march, Eaton and O'Bannon had to deal with mutiny, extremes of hunger and thirst and constant brawling between the diverse elements of their motley expeditionary force. Arriving at Derna in late April 1805, they immediately demanded Yusuf's surrender. When the tyrant declined to hand himself over, they drew up an assault plan to capture his stronghold.

Under cover of three American warships in the harbor, Hamet Bey and his Arabs were to attack Yusuf's castle while the Marines, Greeks and Turks seized the harbor fort. Things did not go according to plan. The naval gunfire made no impression on Yusuf's garrison, and Hamet and his Arabs held back, waiting to see how the situation developed.

Under heavy fire, and with an enemy force approaching, O'Bannon led a charge in which Eaton and three Marines fell wounded. Yusuf's men were driven from the walls of the harbor, the Stars and Stripes was raised

and the unspiked guns turned on the castle. Hamet Bay rallied his Arabs and swept away the last remnants of resistance. Within two hours of the first shots being fired, Derna was in the hands of Eaton's polyglot army. A peace treaty with Tripoli was concluded in June 1805.

Hamet Bey reputedly rewarded O'Bannon with his sword, a weapon of Mameluke design, decorated with an ivory hilt topped with a golden eagle's head. The "Mameluke sword" served as the pattern for the swords still carried by Marine officers on ceremonial occasions.

The War of 1812

In military terms, 1812 was dominated by Napoleon Bonaparte's invasion of Russia. But far from the great European war fronts, a small, bitter struggle was being waged across the Atlantic. The American War of 1812 was ostensibly caused by British violation of American neutrality during British naval operations against the French and their allies. The United States declared war against Britain on 18 July 1812, to "defend the freedom of the seas", but also with an eye on the acquisition of British territory in Canada. On the outbreak of war, the total of US military personnel stood at 12,631, with the Marine Corps comprising 10 officers and 483 enlisted men. There were only three first-class warships in the Navy, *Constitution*, *President* and *United States*. Their task was to harass the British by chasing down commerce and whenever possible to engage single British warships.

In this they were conspicuously successful. *Constitution* destroyed the British *Guerrière* on 19 August off Nova Scotia; on 25 October *United States* seized *Macedonian* off Madeira; and, on 29 December, *Constitution* sank *Java* off Brazil, earning the nickname "Old

Below: British forces, under Major-General Robert Ross, advance on Washington in August 1814.

Ironsides". Marines were involved in all these actions, leading boarding parties and delivering musket fire from the fighting tops. In 1813, Marines participated in the actions which gave Commodore Perry a superiority over the British naval forces on Lake Erie, which he retained until the autumn of 1814 when reverses in other sectors forced the US Navy to relinquish control of the lake.

Once again the war threw up a Marine Corps hero, in the youthful form of a 23-year-old Lieutenant, John Marshall Gamble. On 22 October 1812, Gamble and 31 Marines sailed from the Delaware Cape on board *Essex*, under the command of Captain David Porter. In April 1814, off the Galapagos Islands, Porter intercepted and captured three British whaling ships. One of the captured vessels was then commissioned as the USS *Greenwich* and put under the command of Gamble with a crew of 14 men. Thus Gamble became the only Marine ever to captain a ship of the US Navy.

In July 1813, Gamble displayed great coolness and seamanship when *Greenwich* outmaneuvered and captured the *Seringapatam*, an armed British whaler. In October, Porter left him in charge of three British prize ships at Nukuhiva in the Marquesas Islands, hastily fortified and garrisoned by 22 men. Within an ace of being overwhelmed by mutiny, hostile natives and his British prisoners, Gamble fled for his life aboard one of the prize ships, *Sir Andrew Hammond*, with a crew of seven wounded men.

After a 15-day voyage, Gamble landed in the Sandwich (Hawaiian) Islands, where his ship was given a refit by friendly traders and natives. In return, Gamble undertook a mission to ferry tribute to the chieftain of a nearby island. En route he encountered the British warship *Cherub* and was forced to surrender. Gamble and his men remained prisoners for the rest of the war. The young Lieutenant returned to New York in August 1815 and resumed a distinguished career with the Marine Corps, from which he retired in 1834 with the rank of Lieutenant-Colonel.

Napoleon's abdication in April 1814 released large numbers of British Peninsular War veterans for service in the American theatre. The British mounted three geographically separate operations against the United States. In the north, 14,000 troops, transferred from Europe and led by General Prevost, marched on New York from Montreal at the end of August. Prevost was forced to withdraw after an attempt to secure

the strategically vital Lake Champlain and was defeated in a naval engagement on 11 September.

The British were more successful in the Chesapeake Bay area, where on 19 August 1814 Admiral Cockburn's squadron landed 5,400 veterans of the Peninsular War under General Robert Ross. The British advanced on Washington, scattering 6,000 local militia at Bladensburg, a few miles outside the capital. The militiamen took to their heels at the first rattle of musket fire, but a battalion of 114 Marines, commanded by Captain Samuel Miller, and sailors under the command of Commodore Joshua Barney, stood their ground. They inflicted 250 casualties on the British with their muskets and an 18-pdr cannon, delaying the British advance for two hours.

The British lingered only to burn the Capitol and the White House, although leaving the Marines' barracks untouched, before withdrawing to mount an unsuccessful attack on Baltimore.

Above: The death of General Pakenham during the Battle of New Orleans, 8 January 1815.
Left: The Battle of Lundy's Lane, 25 July 1814.

The Marines gained sweet revenge for the burning of Washington in January 1814. On 13 December 1814, a British expeditionary force of 7,600 Peninsular veterans, under the command of Major-General Sir Edward Pakenham, landed on the Gulf Coast. Their objective was New Orleans. Already well-placed to counter this threat were the militiamen commanded by Andrew Jackson, who had originally been despatched to Alabama in March 1814 to put down an Indian rising. Jackson's force was an extraordinary mixture of fighting men, among them Jean Lafitte's pirates and 300 Marines under the command of Major Daniel Carmick, who was mortally wounded in action against the British on 28 December.

Pakenham's main thrust against Jackson's base at New Orleans began on 8 January 1815, the British advancing in close order to be mown down by the well-dug-in defenders. Over 2,000 British troops were killed or wounded, with Pakenham among the dead. American casualties were fewer than 100. The war, and British violations of American neutrality, was brought to an end by the Treaty of Ghent, signed on 24 December 1814, ironically a month before Pakenham lost his life before the defenses of New Orleans.

Congress expressed its thanks "for the valor and good conduct of Major Daniel Carmick, of the officers, non-commissioned officers, and Marines under his command". The same, alas, cannot be said for Marine Commandant Franklin Wharton, who had fled Washington after the Battle of Bladensburg. Charges of neglect of duty and conduct unbecoming an officer and a gentleman were brought against Wharton by Marine Captain Archibald Henderson, an exemplary fighting man who had commanded the Marine detachment aboard *Constitution*. At a trial in September 1817, Wharton was acquitted, but he died under a cloud.

Wharton's successor, Major Anthony Gale, was another bad lot, lasting only two years before he was court-martialled and dismissed the service. Fortunately for the Marine Corps, he was succeeded by one of its outstanding Commandants, the 38-year-old Archibald Henderson.

At Home and Abroad

Under Henderson's vigorous command, training, discipline and the Marines' personal appearance underwent dramatic improvements. Although the complement of the

Seminole Indians observe a waterborne Marine patrol in the Florida Everglades during the Seminole War. Marine Commandant Henderson was rewarded for his role in the suppression of the Seminoles by promotion to Brigadier-General. He was the first Marine officer to hold general officer rank. For the enlisted Marines of the 1830s and '40s, life was tough. Drunks were forced to drink several quarts of salt water, an extremely unpleasant way of sobering up. Falling asleep on watch was punished by several months of walking guard duty wearing an iron collar and ball-and-chain. Disobeying the commands of a sentry earned 12 lashes with the cat o' nine tails. As late as 1843 a private's pay was only $6 a month, plus $30 a year uniform allowance.

Corps was small, it found employment in a wide range of duties: suppressing piracy in the Caribbean; landing on the Falkland Islands in 1832 to protect American interests in the South Atlantic; and in the same year participating in a punitive expedition in Sumatra after the harassment of American shipping by Malay pirates.

At home the Marines saw service as emergency firefighters — notably during the Boston fire of 1824 — and guardians of public order. They successfully combined both roles when arsonists set fire to the US Treasury in Washington DC. Marines from "Eighth and Eye" — Corps headquarters located at Eighth and I Streets, SE — helped to put out the fire and then stood guard over the nation's treasures.

A milestone in Marine history was reached in 1834 with the passing by Congress of an "Act For the better Organization of the Marine Corps", which frustrated attempts to merge the Corps with the Army and preserved it as a semi-autonomous body within the Naval establishment, ashore and afloat, except when detached for service within the Army by order of the President. It also set the Corps' peacetime strength at 63 officers and 1,224 enlisted men.

Almost immediately, President Jackson detached all able-bodied Marines for service with the Army in the war against the Creek Indians of Georgia and the Seminoles of the Florida Everglades. Commandant Henderson quickly assembled a two-battalion regiment consisting of 38 officers and 424 enlisted men. The story goes that before leaving Marine headquarters, Henderson pinned the following note to his door: "Gone to fight the Indians. Will be back when the war is over. A. Henderson. Col. Comdt." The Marines played an important role in bringing the Creek war to an end in the summer of 1836. The Seminoles proved more intractable, and the campaign against them was brought to an end in 1842 without a treaty having been signed. During these operations against the Indians, the Marine Corps gained field experience which was to prove of immense value in the Mexican War.

The Mexican War

The war of 1846–47 with Mexico sprang from the hectic period of American expansion which followed the Louisiana purchase of 1803. As its frontiers pushed westwards and southwards, the United States came into conflict with Mexico over the latter's territory in California and Texas, which was annexed by the United States in 1845.

Contention over the lands north of the Rio Grande River flared into hostilities on 18 May 1846, when Marine detachments from vessels under the command of Captain John H. Aulick, USN, clashed with Mexican troops at Burrita, 15 miles upstream from the mouth of the Rio Grande.

This skirmish was quickly overshadowed as Marines from the Gulf Coast Squadron began operations to establish a blockade in support of General Zachary Taylor's march south from the Rio Grande. The squadron's senior Marine officer, Captain Alvin Edson, combined all the ships' detachments to form a 200-strong Marine battalion. In October 1846, the battalion, reinforced by sailors and supported by the guns of the squadron, made a series of successful raids against the ports of Frontera and San Juan Bautista.

On 16 November, General Taylor occupied Saltillo. Thereafter he was obliged to assume a defensive posture while General Winfield Scott mounted an invasion of Mexico from Tampico, which on 14 November had been seized by a landing party including Edson's Marines. Vera Cruz was Scott's first important objective.

Following General Taylor's defeat of Santa Anna at Buena Vista on 22/23 February 1847, the focus of the war shifted to central Mexico, where Scott's offensive was about to begin. On 9 March his force made an

Top and centre: Two views of the fighting at Churubusco during the final stages of the advance on Mexico City in the summer of 1847.
Left: The ill-fated Mexican General Santa Anna, who was unequal to the task of halting the American advance on Mexico City.

unopposed landing at Vera Cruz, which surrendered after an 18-day siege. Pushing westwards, Scott routed a Mexican army under Santa Anna at Cerro Gordo on 18 April. However, early in May his advance came to a halt at Puebla. He now had barely 6,000 men under his command, as 4,000 of his original force had returned to the United States on the completion of their agreed period of service.

Commandant Henderson — taking advantage of a 1,000-man increase in the Corps' strength authorized by Congress on 3 March 1847 — immediately began a recruiting drive, and obtained Scott's permission to make good some of his losses with Marines. A battalion of Marines, some 350 strong, joined Scott's forces at Puebla on 6 August and was attached to General John A. Quitman's 4th division under the command of Major Levi Twiggs, a hard-fighting veteran of the campaign against the Seminole Indians. In Scott's rear, Marines of the Gulf Coast Squadron had played an important part in the operations which closed the last remaining ports of entry on Mexico's western coast.

On 7 August, General Scott's reinforced army of 11,000 men began its advance on Mexico City, defeating Santa Anna at Molino del Rey on 8 September. Still outnumbered by at least 2:1, Scott pressed on to clear the last major defensive obstacle before Mexico City. This was Chapultepec, a 200ft-high outcrop topped by a fort. He launched his attack on Chapultepec at 0800 hrs on 13 September. Major Twiggs, armed with a double-barreled shotgun, was killed in the advance to the base of the Mexican fortifications. Under heavy fire, the American troops scaled the walls and seized the fort. The Marines suffered 24 casualties in the attack, and among the wounded was Second Lieutenant C.A. Henderson, the son of the Commandant.

Thrusting on to Mexico City with elements of 4th Division and Marines, General Quitman was held up by fierce enemy resistance. Marine Captain George H. Terrett — a veteran of the 1832 campaign against the Malay pirates — moved his company forward without orders to attack a line of Mexican batteries. He pursued the fleeing gunners and then broke up a counter-attack by Mexican lancers. Mexico City was taken on 14 September. With the city secured, Marine Lieutenant Augustus S. Nicholson cut down the Mexican colors and

Right: A diorama at the Quantico Marine Corps Museum showing a Marine officer urging on his Leathernecks as they overrun the San Cosmo Gate, Mexico City.

ran up the Stars and Stripes. In the Treaty of Guadulupe Hidalgo, signed on 2 February 1848, the United States acquired 500,000 square miles of territory up to the Rio Grande, including California, Nevada, Utah and large tracts of Arizona, New Mexico, Colorado and Wyoming.

In honour of the Marines' contribution to victory, the citizens of Washington DC presented Commandant Henderson with a blue and gold standard bearing the motto: "From Tripoli to the Halls of Montezuma". Later, Marine service in the war was immortalized in the opening line of the Marines' hymn, "From the Halls of Montezuma to the shores of Tripoli".

Prelude to the Civil War

After the Mexican War, the Corps' establishment was cut back to its authorized peacetime strength of 1,224 enlisted men. This did not prevent the Marines from continuing to fulfil a worldwide role. During this period the Corps' expanding range of duties was linked to the United States' growth as a maritime power. Widening American commercial interests overseas frequently required a strong Navy presence to protect the lives and property of its citizens in troublespots torn by revolution or civil strife.

Marines were active in the suppression of

Below: US forces on the march from Puebla, August 1847.

the slave trade; the safeguarding of American interests in Nicaragua, Uruguay, Panama, Argentina, Fiji, Hong Kong and Shanghai; the establishment of diplomatic relations with Japan; and military action in China.

From the mid-nineteenth century, China was torn apart by the Taiping religious war, a conflict which claimed 20 million lives before it ran its 15-year course. Warring factions exploited the crisis to make regular raids on foreign property, and the Marines were frequently called upon to protect American citizens from the mob.

In September 1856 the situation near Canton boiled over when the four "Barrier Forts" on the Pearl River — bristling with 176 cannon and garrisoned by 4,000 troops — opened fire on American ships. A force of Marines, placed under the command of Captain John D. Simms, was conveyed up the Pearl by the Navy steam frigate *Jacinto* and two sloops-of-war, *Portsmouth* and *Levant.*

Simms' attack on the "Barrier Forts" began on 20 November. In three days' hard fighting the 287 Marines and sailors under his command captured the four Chinese strongholds, killed an estimated 500 of their defenders and routed an army of thousands. American losses were seven killed and 20 wounded.

Marines were also required to keep the peace at home. On 1 June 1857, elections in Washington DC were disrupted by an armed gang, the self-styled "Plug-Uglies", from Baltimore. The city police were overwhelmed and Mayor Magruder called for assistance from Marine Headquarters. The Marines and the "Plug-Uglies", now deploying a cannon, came face to face near Fifth and K streets. At that moment, a ramrod-straight, white-haired figure in civilian clothes and carrying a gold-headed cane stepped from the crowd and advanced on the gang from Baltimore. It was the 74-year-old Commandant Henderson, who placed himself squarely in front of the cannon's muzzle and calmly told the mob: "Men, you had better think twice before you fire this piece at the Marines."

One of the bolder "Plug-Uglies" thrust a pistol in Henderson's face, whereupon the Commandant seized the man in a vise-like grip and hauled him away to be placed under arrest. Then a ragged volley of shots was aimed at the Marines, who "poured in an answering fire". The rioters fled.

Two years later, on 6 January 1859, Henderson died while still in office. He had served as Commandant for 39 years, under ten Presidents and during the United States' turbulent early years of nationhood. There can be no better description for him than "The Grand Old Man of the Corps".

The Civil War

In the autumn of 1859, the Marines were faced with a civil disturbance which dramatically reflected the tensions between the Northern and Southern states which were inexorably tearing the Union apart.

On the night of 16 October, the fanatical abolitionist John Brown seized the government arsenal at Harper's Ferry, Virginia, in a bid to raise the slaves in revolt in the Shenandoah Valley. The Navy Department ordered all available Marines in Washington to proceed at once to Harper's Ferry.

The new Commandant, Lieutenant-Colonel John Harris, despatched 86 men by rail from Corps HQ, led by Lieutenant Israel Greene. On the following day they reported to the overall commander of the operation at Harper's Ferry, Colonel Robert E. Lee, USA.

Brown had barricaded himself in the engine house of the arsenal, and on the morning of the 18th Greene was ordered to prepare a storming party. Following Brown's rejection of an ultimatum, delivered by Lieutenant J.E.B. Stuart, USA, Greene's men forced their way into the engine house. As Greene burst through the shattered doorway, Brown fired a shot, killing Marine

Below: Marines parade outside the barracks in Washington in 1861, on the eve of the Civil War.

Union Zouave units engage Confederate cavalry during First Bull Run, a battle noted for the general incompetence displayed by both sides. It was here that a Marine unit broke and ran.

Below right: Union forces cross the Potomac during the Civil War.

Private Luke Quinn. Before Brown could reload his carbine, Greene had cut him down with a sabre slash, which brought an end to all resistance. When Brown had recovered sufficiently from his wounds, Greene and his detachment escorted him to Charles Town. Subsequently he was tried for treason by the State of Virginia and hanged.

Following the election of Abraham Lincoln as President in November 1860, the battle lines were drawn up between the North and the seccessionist Southern States. On 18 February, Jefferson Davis was elected president of the Confederacy. Hostilities began on 12 April when shore batteries from Charleston, South Carolina, fired on Fort Sumter.

During the months preceding the outbreak of war, the divided loyalties felt by many officers had a debilitating effect on the United States' armed forces. As the prospect of secession by the South loomed, many officers from the South, or whose sympathies were strongly in favor of the South, resigned their commissions and offered their services to the Confederacy. The Army lost one-third of its officers (among them Robert E. Lee and J.E.B. Stuart) and the Navy one-sixth. In the Marine Corps, approximately half of its captains resigned, including officers of the caliber of George H. Terrett. Nearly two-thirds of the Corps' lieutenants, and half the second lieutenants, left the service, among them John D. Simms, the hero of the "Barrier Forts", and Israel Greene. With the exception of Major Henry B. Tyler, the field officers of the Corps remained loyal to the Union.

The Corps quickly appointed 38 new officers, but few of them had any previous military experience. In July 1861, Congress authorized an increase of 28 officers and 750 enlisted men for the Marines, raising the strength of the Corps to 3,000. Later in 1861, President Lincoln authorized two increases of 500 men, but at no time during the Civil War did the strength of the Corps exceed 3,900 men. (The Confederate Marine Corps reached a peak strength of 600 men.)

During the course of the war, the Marines' principal function was to serve as detachments on board ships of the Navy. The chances for action ashore were limited and confined to small bodies of men. Marines who served ashore did so either as part of a ship's landing force or directly assigned to the Union Army. Although the Civil War saw many gallant acts by Marines, it was not the Corps' finest hour.

An ominous note was struck at the First Battle of Bull Run on 21 July 1861, when an inexperienced Marine unit under the command of Major John G. Reynolds broke and ran after a series of attacks by Confederate riflemen and cavalry commanded by Colonel J.E.B. Stuart. Any criticism of these raw troops must be tempered by the general incompetence displayed by both sides at Bull Run, a battle cuttingly described by the German commander Helmuth von Moltke as little more than "two armed mobs pursuing each other".

Clearly the Marines were better suited to the amphibious operations conducted along the Southern coast as part of the Union's strategy of blockading the Confederacy's ports. In August 1861 Marines attached to the ships *Minnesota*, *Cumberland*, *Susquehanna* and *Wabash* took part in the

capture of Fort Clark and Fort Hatteras in North Carolina, closing the Hatteras Inlet to British blockade runners and Confederate privateers, who were using it as a base from which to harry Union shipping. In October–November 1861, a 300-man Marine battalion, under the command of Major John G. Reynolds, was earmarked for an amphibious operation against Port Royal, South Carolina, planned by Naval Flag Officer Samuel F. DuPont, commander of the South Atlantic Blockade Squadron. However, the Marines took no part in the action as their unseaworthy transport, the sidewheeler *Governor*, foundered in rough weather on the way to Port Royal. After two more equally abortive missions, the amphibious battalion was disbanded and the men assigned to their ships' detachments.

On 11 May 1862 the scuttling of the Confederate warship *Virginia* (the refitted

Above: Petersburg, Virginia, an important supply base for General Robert E. Lee's Confederate Army of North Virginia. At the end of the Civil War it endured a 10-month siege. Lee's defence was a tactical masterpiece, and he was forced to abandon Petersburg only a week before the surrender of the Army of North Virginia at Appomattox on 9 April 1865.

Merrimack) during the evacuation of Norfolk opened up the James River. Four days later Union warships approached Drewry's Bluff on the James, about eight miles from Richmond. Here they came under fire from Confederate shore batteries stationed on the bluff. The *Galena* was returning fire when a direct hit caused a violent explosion. Marine Corporal John Mackie rallied the survivors, carried off the dead and wounded and got three of the ship's guns back in action. These acts of heroism made him the first Marine to win the Medal of Honor. By the end of the war he had been emulated by 17 of his fellow Marines.

Nevertheless, frustration rather than derring-do was the keynote of the Marines' war. In August 1863, Marines under the command of Major Zeilin were tasked with assisting Union forces in the attack on Fort Wagner, South Carolina. Fired with enthusiasm, Zeilin attempted (unsuccessfully) to expand his 300-strong battalion into a regiment and instil in his men a set of basic amphibious tactics. The results were not encouraging, and Zeilin's Marines were left to kick their heels. Zeilin was replaced by Lieutenant-Colonel John Reynolds, who reduced the regiment to a battalion and turned it over to Captain Charles G. McCawley. In September, Marines were attached to the expeditionary force assembled by Rear-Admiral John A. Dahlgren for an attack on Fort Sumter. The assault was a dismal failure; 44 of the Marine party of 150 men were killed, captured or wounded. The battalion was taken out of combat for rest and recuperation, but suffered further losses from disease and was broken up in 1864.

The demoralized Commandant, John Harris, died on 12 May and was succeeded by Major Zeilin. On the previous day, Secretary of State Gideon Welles had retired all Marine officers who ranked senior to the new Commandant.

By the end of 1864 the only Southern port which continued to defy the Union was Wilmington, North Carolina. Entrance to the port was by the appropriately named Cape Fear River, guarded by the 44 large-caliber cannon of Fort Fisher and over 1,000 troops.

The first attempt to take Fort Fisher was a fiasco engineered by a bumbling amateur soldier, Major-General Benjamin F. Butler, a politician from Massachusetts. The Union's General-in-Chief, Ulysses S. Grant, replaced Butler with a professional, General A.H. Terry, leading a force of 8,500 men under the overall command of Rear-Admiral David D. Porter.

Porter devised a plan of attack which bordered on the reckless. Following a "softening-up" bombardment from his warships, 1,600 of Porter's seamen and 400 Marines would make a frontal attack on Fort Fisher while Terry's troops launched the main thrust from the rear. Throwing seamen armed with cutlasses and pistols — and covered only by Marine riflemen — against the walls of a heavily defended fort displayed an optimism which one of Porter's subordinates characterised as "sheer murderous madness".

Amid scenes of great confusion, the seamen and the Marines were landed before Fort Fisher on the afternoon of 15 January 1865. The Marines arrived too late to support the bluejackets, who were mown down by the fort's defenders. Ordered into the attack themselves, they suffered a similar fate. Combined casualties were 309 dead and wounded. Nevertheless, the diversion served its purpose and Terry's troops stormed Fort Fisher from the rear. But the bloody price of victory triggered off a furious controversy over Porter's tactics, in which the Admiral defended himself with the improbable assertion that, "Had the covering party of Marines performed their duty, every one of the enemy would have been killed".

The furore over the attack on Fort Fisher did little to promote the cause of the Marines as an independent force specifically trained and equipped to undertake amphibious operations. The Civil War came to an end on 9 April 1865 without the Corps having found a basic role which would dispel the clouds still hanging over its future. Moreover, its marginal influence on the course of the war can be gauged from its casualty figures of 148 killed in combat and 312 dead from other causes. The combined losses of North and South from all causes are estimated at approximately 520,000 men.

Left: Major-General Benjamin F. Butler, the incompetent amateur soldier who bungled the first attempt to take Fort Fisher.

Below: The bombardment and assault on Fort Fisher, 15 January 1865. The high casualties suffered by the Marines and bluejackets in the action led to a political furore which did little to enhance the reputation of the Corps.

From Korea to the Philippines

The Marine Corps emerged from the Civil War with its morale and reputation at a low ebb. In the summer of 1866, the House of Representatives considered a resolution for the abolition of the Corps and its transfer to the Army. Marine Commandant Zeilin mounted a fierce political counter-offensive, enlisting support from the Navy. In February 1867 the resolution was thrown out by the House Naval Affairs Committee, which recommended that the Marines' "organisation as a separate corps should be preserved and strengthened . . . [and] that its commanding officer shall hold the rank of brigadier-general".

Service at sea continued to be the primary role of the Marines, although the advent of steam ships meant that they were no longer stationed as sharpshooters on the fighting tops. The Marines serving on foreign sea duty after the Civil War were soon re-acquainting themselves with far-flung parts of the world in the role of peacekeepers and protectors of American lives and property: in the 1860s, small-scale landings were made in Formosa, Japan and Uruguay; in the 1870s Marines went ashore in Mexico, Colombia and the Hawaiian islands; in the 1880s it was the turn of Egypt, Korea, Haiti, Samoa and the Hawaiian islands; followed in the 1890s by

Right: A contemporary sketch map of the American operations in Korea in 1871, including the positions taken by the Marines and bluejackets.
Opposite, top: Marine headgear of the late 1870s. Clockwise from top left: field officer, full dress; company officer, fatigues; all officers, undress pattern; full dress, enlisted men; fatigue, enlisted men; undress, bandsmen; full dress, company officer.
Opposite, bottom: A Marine racing crew pose proudly for the camera aboard USS *Olympia*.

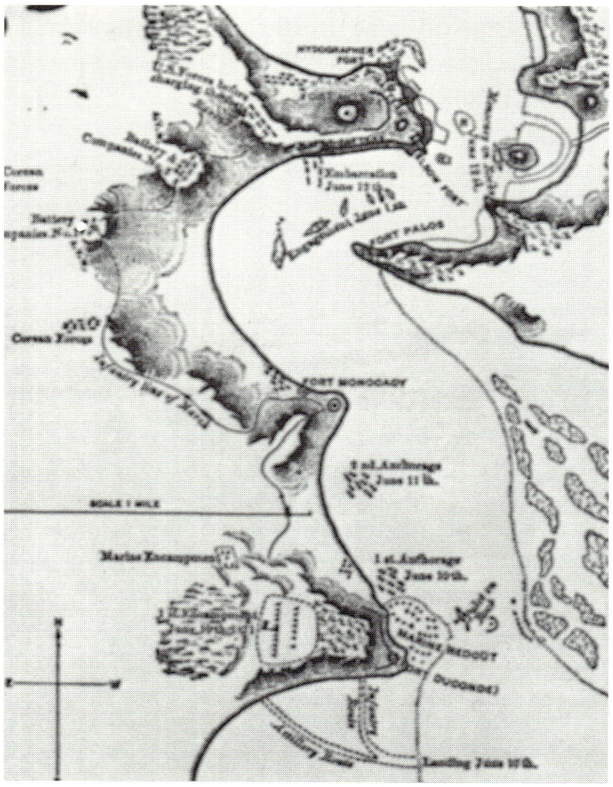

Argentina, Chile, Korea, North China, the isthmus of Panama and Nicaragua.

In most cases, the mere presence of the Marines was sufficient to counter a threat to American interests. But there were exceptions. In the summer of 1871, the American Minister in China, Frederick Low, was despatched on a peace mission to Korea, whose people had already displayed their xenophobia by massacring the crew of an American ship, *General Sherman*, which had gone aground in the Han River.

Low arrived on the west coast of Korea in the latter part of May on board the *Colorado*, flagship of the Asiatic Fleet, accompanied by four other ships under the overall command of Rear-Admiral John Rodgers. Arriving off the mouth of the Han River, Rodgers sent a small scouting party upstream to find an approach to the Korean capital, Seoul. As it worked its way up the channel, the party came under fire from one of five forts which guarded the lower reaches of the Han River.

The Koreans showed no inclination to apologize for this incident, and ten days later Rodgers and Low set about mounting a punitive expedition. A force of bluejackets and a Marine brigade of four officers and 105 enlisted men, led by Marine Captain McLane Tilton, were ordered to capture the Korean forts. This was made no easier by the Marines' armament of obsolescent breech-loading muskets. In contrast, the garrisons of the Han River forts were armed with modern repeaters.

The operation began on 10 June with an assault across a mud flat against the smallest of the forts. Cowed by a naval bombardment, the defenders offered little resistance. A second fort was captured with equal ease on the following day. Tilton then turned his attention to the heavily fortified "Citadel", which had to be approached over mountainous terrain which concealed Korean riflemen. The final assault was pressed home under heavy fire up a 150-ft hillside. After a fierce hand-to-hand fight, the enemy flag was torn down by Marine Corporal Charles Brown and Private Hugh Purvis, both of whom were subsequently awarded the Medal of Honor. Four more Medals of Honor were awarded for this action, in which the Marines accounted for over 200 Koreans at a cost of only two casualties. Their opponents must have been very poor marksmen. Thereafter the Koreans prudently refrained from hostile acts against Americans.

On 11 July 1882, the British fleet bombarded the Egyptian city of Alexandria after a serious outbreak of rioting by *fellah* mobs against foreign influence in the region. With large sections of Alexandria ablaze, 60 blue jackets and 73 Marines from the European Squadron, under the Command of Captain H.C. Cochrane, were landed to assist in restoring order, prevent the spread of fires and protect the American Consulate, which had thus far escaped serious damage. Cochrane, a brilliant officer and legendary martinet, quickly swept through the streets around the Consulate, establishing a riot-free zone and earning these words of praise from the British commander Lord Charles Beresford, whose arrival with a force of 4,000 men had brought the uprising to an end: "To your smart, faithful force great credit is due . . . I have represented these facts to my government."

Much nearer home, the Marines were called in to restore order in the strategically sentitive region of Panama when in 1885 its population rebelled against its Colombian rulers. In April of that year two battalions of Marines were sent to Panama City and formed into a brigade under Colonel Charles Heywood, the first time in its history that the Marines had operated as a brigade. By the last week in April the Marines had secured and maintained military control over

Panama City, handing over to Colombian regulars on 30 April. The Marines had now gained a well-earned reputation for the efficient execution of such operations. When the captain of a British warship at anchor in Panama City's harbor was informed of the arrival of the Marines, he said, "Tranquillity is then assured".

The enforced tranquillity of Stateside garrison duty was often relieved by such law-and-order assignments as assisting the authorities in the destruction of illegal distilleries and the enforcing of the revenue laws. In the spring of 1877, two battalions of Marines assisted the Army in suppressing labor riots in Baltimore and Philiadelphia which flared up during a major railroad strike.

Throughout this period, the eighth Commandant of the Corps, Charles G. McCawley, worked hard to improve its status. McCawley supervized the drawing-up of the first standardized table of organization for the Corps; introduced the first mass-produced uniforms; reorganized and reinvigorated the Headquarters Marine Corps cadre; and improved career prospects for non-commissioned and enlisted men. He also succeeded in drawing Marine Corps officers from the ranks of Naval Academy graduates; between 1881 and 1897 all 50 officers who entered the Corps were Academy graduates.

One of McCawley's less popular measures was his ban on the sale of beer in Marine canteens, and his somewhat quixotic encouragement of a group calling itself the Marine Temperance Union. McCawley also provided an imprtant footnote to musical history when in 1880 he appointed 26-year-old John Philip Sousa as leader of the Marine Band. Under Sousa's inspiring leadership, the Band became one of the world's finest musical groups.

McCawley's reforming work was continued by Colonel Charles Heywood, who succeeded him as Commandant in 1891. Heywood took an important step in the further professionalization of the Corps, creating the School of Application in Washington, fore-runner of the Marines Basic Schools system. Under his leadership, the number of Marine bases increased from 12 to 21; one of them, located at Port Royal, South Carolina, was to become Parris Island, the most important training area for Marines on the eastern seaboard. In training, Heywood placed great emphasis on target practice and

Top: Marines and seamen drawn up for inspection aboard USS *Essex* in 1888. *Above:* Marines at bayonet drill, c. 1890. In the background is the receiving ship USS *New Hampshire*.

marksmanship. In the wider political sphere, he skilfully preserved the integrity of the Corps against renewed pressure to combine the Marines with the Army's regiments of artillery.

War with Spain

The Spanish—American War of 1898 was once described as a "glorious national picnic", which nevertheless signalled the emergence of the United States as an imperial power. The immediate *casus belli* was an incident in which the US battleship *Maine* exploded and sank in Havana Harbor, Cuba, on 15 February 1898, with the loss of 238 sailors and 28 Marines. (In all probability, the explosion was caused by unstable ammunition.)

American public opinion — cynically manipulated by the newspaper baron William Randolph Hearst — was already incensed by the brutal manner in which the Spanish had put down the Cuban insurrection of 1895–8, and war was declared on 25 April on a floodtide of emotion.

There were two theaters of war: the Spanish possessions in the Caribbean (Cuba and Puerto Rico) and the Philippines. On 1 May, Commodore George Dewey's Asiatic Squadron annihilated Admiral Montojo's fleet in Manila Bay. On 3 May, Dewey landed a contingent of Marines from the USS *Baltimore*, under the command of Lieutenant Dion Williams, to occupy the Spanish naval station at Cavite. The Marines were the first to land on Spanish territory and raise the American flag. Manila capitulated after the arrival of American Army troops at the end of June. Nine days earlier, Marines from Charleston had landed unopposed on the Spanish island of Guam. The Pacific phase of the war was over.

At home the Marines were preparing for a landing on Cuba. On 16 April, Commandant Heywood had been instructed to organize a Marine battalion to serve in Cuba. With great drive and efficiency, the Marines assembled 647 men, organized into four infantry companies and an artillery battery, under the command of Lieutenant-Colonel Robert W. Huntington. Heywood personally oversaw the equipping of the force: "... [the battalion] was supplied with all the equipment and necessities for field service under conditions prevailing in Cuba, including mosquito netting, woolen and linen clothing, heavy and light weight underwear, three months' supply of provisions, wheelbarrows, pushcarts, pick-axes, shovels, and a full supply of medical stores".

Within a few weeks the Marine battalion had been transported by the *Panther*, commanded by Commodore G.C. Reiter, to Key West in Florida, there to await orders.

It soon became clear that Commander Schley's Flying Squadron needed a shore base to support the blockade of the Spanish fleet in Santiago. Huntington's Marines were ordered to seize Guantanamo Bay, about 40 miles east of Santiago, as a coaling station. Intelligence reports indicated that there were about 8,000 Spanish troops in the area. On the afternoon of 10 June, Huntington's battalion was landed from *Panther*, becoming the first American troops to establish a beach-head on Cuban soil.

The landing was unopposed, which was just as well as the prickly Commodore Reiter initially refused to send ashore the Marines' small-arms ammunition, on the grounds that he needed it as ballast for his ship. He was quickly brought into line by the senior Naval officer in the area, Commander Bowman McCalla, captain of the cruiser *Marblehead*, who ordered the immediate landing of 50,000 rounds of small-arms ammunition. The grateful Marines named their beach-head

Above: The wreck of the USS *Maine* in Havana habor, Cuba. Twenty-eight Marines died in the explosion, which is now thought to have been caused by the spontaneous detonation of unstable ammunition. But the Spanish took the blame, and in the ensuing war the United States took the Philippines, the Pacific Island of Guam and Peurto Rico.

Left: Commodore George Dewey, whose Asiatic Squadron annihilated the Spanish fleet in Manila Bay.

Above: The Battle of Manila Bay.

Below: Marine riflemen ready themselves for an enemy counter-attack after the unopposed landing at Guantanamo Bay on 10 June 1898.

bivouac Camp McCalla.

On the afternoon of 11 June, the Spanish roused themselves from their torpor, attacking a Marine outpost and killing two men. After three days of desultory fighting, Huntington decided to cut off the Spanish water-supply, which was located at a fortified well in the village of Cuzco, six miles south-west of Camp McCalla. On the morning of 15 June, two rifle companies, under the command of Captain George F. Elliott (later the 10th Commandant of the Marine Corps) and 50 Cubans marched on Cuzco. Supporting fire for the attack was provided by the warship *Dolphin*. However, its salvoes began to creep dangerously close to the Marines. Disaster was averted by the initiative and coolness of Marine Sergeant John H. Quick, who improvised a semaphore flag and, with enemy fire whistling about his ears, redirected the *Dolphin*'s fire on to the enemy position. The author Stephen Crane was at Cuzco as a war correspondent and left this memorable thumbnail sketch of the gallant Quick: "I saw Quick betray only one sign of emotion. As he swung his clumsy flag to and fro, an end of it once caught on a cactus plant. He looked annoyed."

Victory in the action at Cuzco dealt a severe blow to the already fragile Spanish morale. On 22 June, 17,000 US Army troops under General Shafter landed at Santiago, defeating General Linares' 35,000-strong army at San Juan on 1 July. Two days later, Admiral Cervera's attempt to break out of Santiago harbor was thwarted, and the city surrendered on 17 July. Fighting continued in Puerto Rico, where General Miles had landed at the head of an expeditionary force on 4 August. On 10 December, hostilities were brought to an end by the Treaty of Paris. Spain relinquished sovereignty over Cuba and ceded Puerto Rico and the Pacific island of Guam to the United States, which also acquired the Philippines at the knock-down price of $20 million.

The war was a turning point for the Marine Corps. The United States had gained a quick victory, but most "picnics" are rather better organized than the American campaign in 1898, which was notable for its chaotic staff work. Army troops sweltered in the tropical

Above: The first picture taken of Marines — and feline mascot — on the island of Guam.

heat in woolen uniforms; at least half of them contracted yellow fever and enteric disease. The Marines, however, were a shining exception — well-organized, well-prepared, quick to respond when the call came, and considerably fitter than their Army colleagues. Only two percent of the Marines personnel suffered from enteric fever, and none of them contracted yellow fever.

On 3 March 1899, Congress approved a bill calling for a permanent Marine Corps consisting of 201 officers and 6,062 men, a 100 percent increase over its strength in 1896. In the longer term, the success of the operation at Guantanamo encouraged the development of amphibious tactics and a detailed consideration of the Corps' role in seizing and holding advanced bases.

As the newly-formed Navy Board acknowledged in 1900, the Marine Corps was "best adapted and most available for immediate and sudden call" in such operations. It directed the Corps to select the personnel and develop the techniques for this task, and by 1902 an Advanced Base battalion was in training.

In July 1910 an Advanced Base school was established at New London, Connecticut (it was moved to Philadelphia a year later). Although principally aimed at the training of officers in the Advanced Base concept, its first class consisted of 40 enlisted men. By 1914 the Marines' combat-ready element was centered on the Advanced Base Force, a brigade-sized combined-arms formation whose prime function was to assault and defend bases for the US Navy.

Before the creation of the Advanced Base Force, each Marine barracks had its own "rookie squad" which gave rudimentary training to new recruits. However, the Advanced Base Force needed men with a standard level of training, and the solution was provided by the setting-up of small Marine recruit depots at the larger Naval stations. The search for greater economy and efficiency soon led the Corps to consolidate all its recruit training at two bases, one on each coast. The Marine Barracks, Parris Island, South Carolina, established its attached

Above: A Marine firing line in action against insurgents in the Philippines in 1901.

recruit depot in 1915. A similar depot was set up at Mare Island, California; the latter moved to San Diego in 1923.

In 1914 the Advanced Base concept was tested when trouble flared with Mexico after Washington learned that Germany was about to supply arms to the insurgent General Huerta. On 21 April, President Wilson ordered naval forces to land and seize the customs house at Vera Cruz. Sailors and Marines went ashore that day, and by 24 April Vera Cruz had been pacified. On the 28th, Army troops occupied the area. They were supported by a Marine brigade of 3,000 officers and men, which remained in Vera Cruz, under Army command, until November 1914.

"He Served on Samar"

The Marine Corps' involvement with the Philippines did not end with the war of 1898. By 1900, Filipino agitation for independence led to the stationing in the Philippines of four Marine battalions, comprising two rifle regiments and two artillery companies, a total of 58 officers and 1,547 enlisted men.

The Marines' service in the Philippines was characterized by long periods of boredom, familiar to everyone who has endured garrison duty, and short spells of savage fighting. Among the Corps' fiercest adversaries were the Moros of Samar Island. In September 1901 450 Moros had surprised and wiped out US 9th Infantry's C Company as it sat down to breakfast, making off with all their rifles and 28,000 rounds of ammunition.

On 24 October 1901, a battalion of Marines landed on Samar, under the command of a brilliant, peppery officer, Major Littleton W.T. Waller. In a month of hard fighting — during which no quarter was given by Waller's men — the Moros were driven deep

Below: Marines patrol the tropical jungle near Olongapo, Luzon, during operations against Filipino guerrillas in 1900.

into the jungle and then overwhelmed in their stronghold in the cliffs of the Basey River.

With peace restored on Samar, General Jacob M. Smith, USA, ordered Major Waller to reconnoiter a telegraph route from Basey to Lanang. Waller's 52-mile march through the jungle began on 28 December, and quickly turned into a nightmare. Boats and provisions were lost in treacherous rivers; bearers mutinied; ten of the 50 Marines died of to fever and exhaustion, and one of them went mad. After emerging from the bush on 15 January 1902, Waller set up a drum-head court-martial, which tried and then executed 11 of the Filipino bearers, a drastic disciplinary measure which resulted in Waller being hauled before an Army court-martial on 11 counts of murder. Waller was eventually acquitted, but a shadow was cast over his subsequent career and probably prevented him from becoming Commandant of the Corps. Nevertheless, his jungle march became a Marine Corps legend. Whenever one of Waller's men was present at a brigade mess, the toast would be raised: "Stand, gentlemen. He served on Samar".

From China to the Caribbean

In the summer of 1900, the Boxer Rebellion burst upon China. Much of its ferocious energy was channeled against foreign influence in China, and the legations of the Western nations, alarmed for their physical safety, requested military protection from their respective governments.

Marines were quickly on the spot. Late on the night of 29 May, a detachment of Marines and sailors from the US warships *Newark* and *Oregon* arrived in Tientsin, some 80 miles south of Peking. Among those who watched them arrive was a young engineer named Herbert Hoover, who later recalled: "I do not remember a more satisfying musical performance than the bugles of the American Marines entering the settlement playing 'There'll be a Hot Time in the Old Town Tonight'."

The Marines were shortly followed by British, Austrian, German, French, Italian, Japanese and Russian landing forces. Two days later a Marine detachment led by Captain John T. ("Handsome Jack") Myers and Captain Newt H. Hall, formed part of the international force which set out by train from Tientsin for Peking, where they marched to the American Legation through dense crowds of silent and hostile Chinese.

On 5 June the rail lines between Tientsin and Peking were torn up by the Boxers. Five days later an international relief force, commanded by Vice-Admiral Sir Edward Seymour, RN, set out to repair the rail line and reach Peking. Harassing attacks by the Boxers brought Seymour's force to a halt 25 miles short of Peking and then forced it back to the outskirts of Tientsin, where it was bottled up in the Hsi-ku arsenal. At the end of June a fresh international force, including 138 Marines commanded by Major Waller, lifted the siege of the arsenal and then joined Seymour in crushing the Boxers in the area,

Below: Members of the International Relief Force pose for the camera in Peking. They include Japanese and British infantry, French Chasseurs Alpin, men of the German East Asia Division, Russian Cossacks, men of the British Royal Navy and Marines, Russian sailors and — seated on the right — US Marines.

Top: A stylized contemporary rendering of the street fighting in Peking between rampaging Boxers and appropriately jut-jawed Marines.
Above: A battalion of Marines parades in Peking's Forbidden City after the defeat of the Boxers.

which was pacified by the end of July.

In Peking, the siege of the legations began on 20 June. At the beginning of August an international force of approximately 18,000 men — including a fresh regiment of Marines under Major William P. Biddle — began its march on Peking, eliminating hostile Chinese forces on the way. They reached the outskirts of Peking on 13 August and raised the siege of the legations on the same day. On the 14th the Allied force made its final assault on the Imperial City, driving the Boxers out. By the end of the month all organized resistance had collapsed. The Marines remained in Peking until the end of September, when they returned to the Philippines.

In the years leading up to World War I, the Marines became increasingly embroiled in the turbulent politics of the Caribbean and Central America, both of which were, and remain, vital to the hemispheric interests of the United States. In the winter of 1903, Marines landed in Panama, whose people were in open revolt against their Colombian rulers. The stakes were high, as the United States was in the middle of negotiating with Colombia for the right to dig a canal across the isthmus of its Panamian province. The Marines restored order and forestalled intervention by Colombian troops. In the words of President Theodore Roosevelt, who had personally ordered the operation: "I took the Canal zone and let Congress debate, and while the debate goes on, the Canal does also." Things were simpler in the days before Watergate and Irangate. The United States quicky recognized Panamanian independence, the price of which was at treaty with the US government which gave it the sovereign right to a canal zone 10 miles wide and the right to keep order within Panama.

When mob violence erupted in Cuba in

August 1906, following fraudulent elections, two battalions of Marines were despatched to assist President Tomas Estrada Palma in restoring order. Within a few weeks there was an entire brigade of Marines in Cuba, totalling 97 officers and 2,795 enlisted men. Shortly afterwards, President Roosevelt ordered the formation of the Army of Cuban Pacification, in which Marines served under the US Army's jurisdiction, remaining in Cuba until 1909. Renewed outbreaks of unrest in May 1912 resulted in the despatch of 1st Provisional Regiment of Marines to Cuba, reinforced by a second regiment which arrived at Guantanamo at the beginning of June. Under the command of Colonel Lincoln Karmany, a provisional brigade was formed which restored order in 26 towns in the Santiago and Guantanamo region until the situation was stabilized in July. Thereafter they handed over their responsibilities to the civilian authorities, pulled back to Guantanamo and then sailed for the United States.

Similar operations were undertaken in Nicaragua, following the outbreak of revolution in 1909 and civil war in 1912. High on the list of Marine priorities was the protection of the substantial American fruit, lumber and mining interests in Nicaragua. In 1912 a Marine regiment was sent to the Dominican Republic — whose economy was wholly dependent on the United States — when political unrest threatened to topple the government. On this occasion, the Marines' presence in their ships in the harbor was sufficient to restore order, but in 1916 a civil war led to the stationing of a brigade of Marines in the Dominican Republic, where they remained until 1925.

In neighbouring Haiti, what passed for politics was a blood-drenched *danse macabre*, which in July 1915 erupted into a sanguinary paroxysm disgusting even by Haiti's grisly standards. On 28 July, a detachment of Marines and sailors from the USS *Washington* went ashore at the Haitian capital, Port au Prince. The next day they were reinforced by a company of Marines rushed from Cuba. Later the Marine force on Haiti was augmented by 2nd Marine Regiment, 1st Marine Regiment and Headquarters, 1st Marine Brigade, the latter commanded by Colonel Waller. Serving with brigade headquarters was a young Marine Lieutenant, Alexander Archer Vandegrift.

In a three-month campaign, the Marines dealt with the principal troublemakers on Haiti, a group of lawless guerrillas known as the Cacos. Once a relative degree of order had been restored, the Marines on Haiti devoted their energies to training Haitian officials in the business of government and directing such civic works as road-building, communications and education. The gendarmerie was staffed and trained by Marines and was gradually transformed into an all-native Haitian force. The Marines were to remain in Haiti until August 1934.

Left: President Theodore Roosevelt, who in 1908 ordered Marine detachments to be removed from ships of the US Navy and threatened to reduce the Corps to little more than a naval station guard. Congress stepped in to preserve the Corps' integrity, much to the fury of the mercurial Roosevelt.

Above: Marines come ashore at Port-au-Prince in the early stages of the Corps' involvement with the chaotic fortunes of Haiti.

2:RETREAT, HELL! WE JUST GOT HERE!
World War One

On 1 June 1916, a small combined force of US Marines and sailors went into action against Dominican insurgents in the town of Puerto Plata. At the same time on the other side of the Atlantic, the British Army was suffering the greatest one-day loss in its entire history — the 60,000 casualties it sustained on the first day of the Battle of the Somme.

The Somme became a symbol of the new kind of war which held Europe in a paralysing grip. On the Western Front, strategy was placed in a straitjacket by the domination of "bullet, spade and barbed wire". Movement had been replaced by mud and the machine-gun. In 1916, an average of 56,000 British soldiers were dying every month.

President Wilson had been elected in 1916 on the slogan "He kept us out of the war". But German unrestricted submarine warfare, and an attempt to push Mexico into action against the United States, forced the President's hand. On 6 April 1917 the United States entered the war as an "associated power", a status which gave her armies a measure of independence from the French and British high commands.

However, in 1917 the US Army was tiny in comparison with the millions locked in combat in Europe. In A.J.P. Taylor's memorable phrase, America's declaration of war was only a "promissory note for the future". It brought handicaps rather than immediate help to the Allies. Tens of thousands of men would have to be conscripted and trained. New munitions factories would have to be built. Tanks, artillery, and even rifles, would have to be provided by the British and French. The Allies would have to hold on until the United States could bring its enormous economic resources to bear and make its presence felt on the battlefields of France.

In April 1917 the Marine Corps strength was 462 officers, 49 warrant officers and 13,213 enlisted men. By December 1918, it had risen to 75,000 officers and men (including the newly-formed Marine Corps Reserve), of whom some 30,000 saw service on the Western Front. Statistically a drop in the ocean, they nevertheless played a part

Left: Marines advancing through a wood shredded by shellfire during the Meuse-Argonne offensive in 1918.

out of all proportion to their numbers.

In June 1917 the first US troops were shipped to Europe under General John J. Pershing, commander of the American Expeditionary Force. Among them were 5th Regiment of Marines, composed of 70 officers and 2,689 enlisted men, completely organized and ready for active service. They made up one-fifth of the first American force to land.

Having arrived in France, 5th Marines were forced to champ at the bit behind the lines, broken up to perform labor and guard duties while inferior units were sent up to the front. This enforced inactivity was in part due to Pershing's doubts about the Marines' ability to adapt to conventional infantry operations, a scepticism which was soon to prove unfounded.

By February 1918, 6th Regiment of Marines and 6th Machine-Gun Battalion had arrived in France. They combined with 5th Marines to form 4th Brigade of Marines, part of 2nd Infantry Division, with a strength of 258 officers and 8,211 enlisted men. They were later joined by 5th Brigade of Marines, with a similar strength.

In March 1918 the Marines were introduced to the front line, in a quiet sector south of Verdun on the Meuse River. They were pulled out for further training after 53 days, having suffered 872 casualties. It was not long before they were thrown into the cauldron of war.

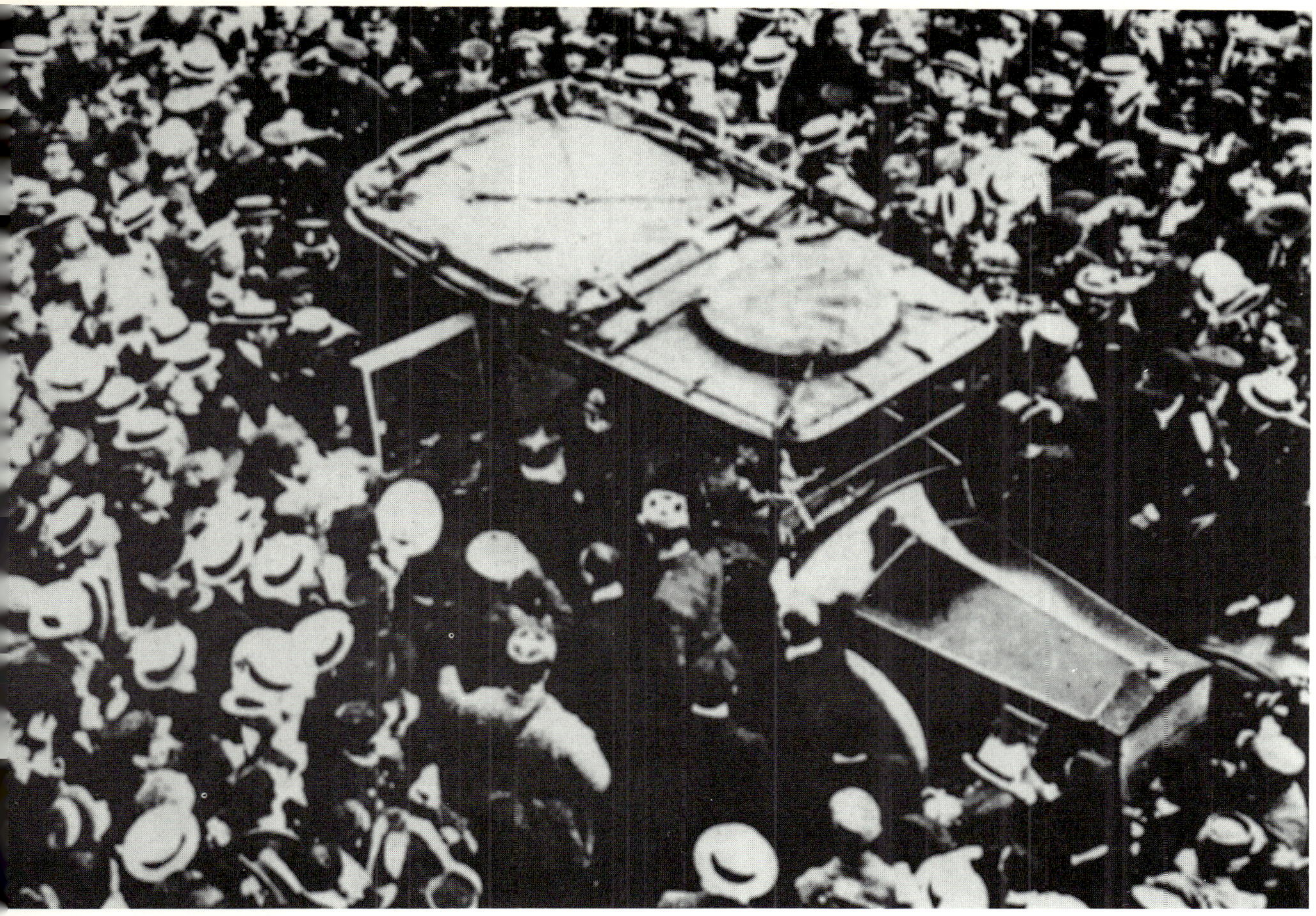

Left: General Pershing, commander of the American Expeditionary force, is welcomed by enthusiastic French crowds.
Far left, top: A vivid recruiting poster with a famous call to arms.
Far left, bottom: They answered the call: applicants arriving at the recruit depot, Parris Island, South Carolina, in 1917.

In the winter of 1917, Germany's warlords, Field-Marshal Hindenburg and General Ludendorff, decided that they must clinch victory before American intervention irrevocably tipped the balance against them. In March 1918 Ludendorff launched a series of massive offensives: on the Somme, penetrating to within 75 miles of Paris; in the Lys–Aisne sector in April–May; and in the Noyon–Marne sector in June–July. On 27 May, 14 German divisions broke through on the Aisne, achieving an advance of ten miles in a day, a distance unequalled since the first weeks of the war in August 1914. Thirty-six hours later the Germans had reached the Meuse at Château-Thierry and were within 40 miles of Paris.

French reserves thrown in to stem the German tide were reinforced by the US 2nd and 3rd Divisions. The sector assigned to 4th Brigade of Marines was Belleau Wood, three miles north of Château-Thierry and at the tip of the German advance. Here 1,200 troops of the battle-hardened 461st Imperial German Infantry had established a formidable defensive position in a square mile of rocky woodland.

The fight for Belleau Wood lasted 20 days. The Marines quickly earned the respect of the Germans with the quality of their marksmanship at ranges of up to 800yds. But much of the fighting was at close quarters as the Marines attacked and the Germans counter-attacked through the stony outcrops and shattered tree-stumps of Belleau Wood. The final enemy strongpoints were neutralized by 3rd Battalion, 5th Marines, on 24 June. The casualty lists reflected the fierceness of the fighting. Two attacks by 1st Battalion, 5th Marines, on 6 June resulted in nearly 1,500 Marines dead or wounded, losses in a day's fighting not exceeded by the Marines until the battle for Tarawa in 1943. In honor of the Corps, the French renamed the wood "Bois de la Brigade de Marines".

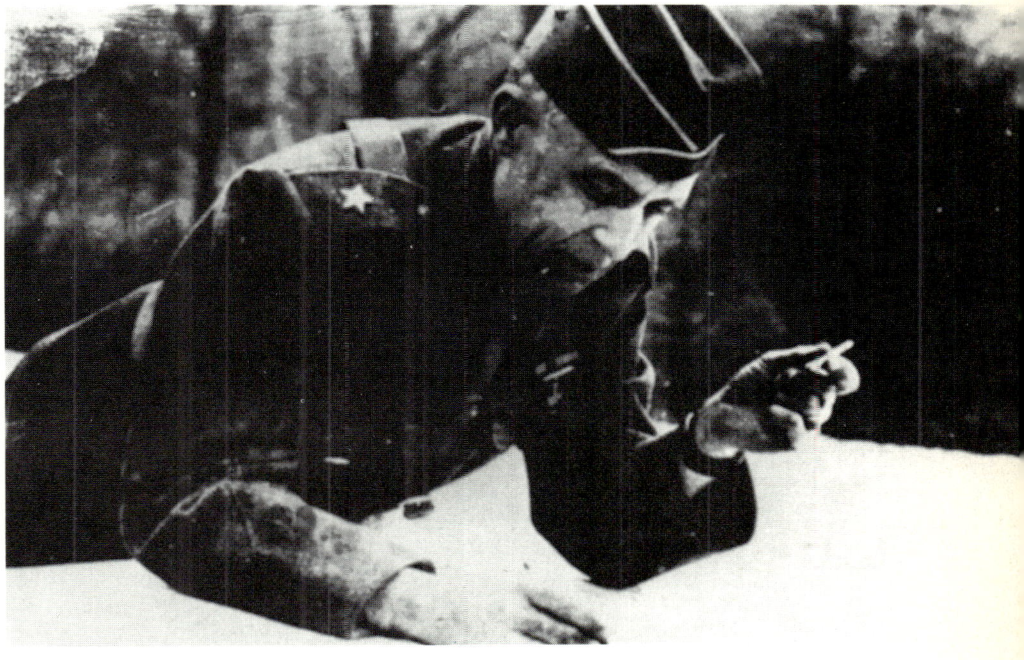

Above: Marine Major-General John A. Lejeune, who assumed command of US 2nd Division in July 1918, becoming the first Marine officer to command an Army division. After World War One, Lejeune became a distinguished Commandant of the Marine Corps.

Above: Marines pause in a French village on their way up to the front at the end of May 1918. They were shortly to face their first major test at Belleau Wood.
Opposite, top: Figures in a bleak landscape; Marine machine-gunners on maneuvers in the winter of 1917-18.
Opposite bottom: Marines man 3-in guns aboard the battleship Pennsylvania in World War One.

More significantly, the action at Belleau Wood, and 2nd Division's counter-attack at Château-Thierry, had halted the German drive on Paris and provided a vital moral tonic for the war-weary British and French. It also marked an important moment in the Marine Corps' history, demonstrating its effectiveness when employed in numbers against seasoned infantry. Above all, it provided the Marines with an arena in which to display the daring and *esprit de corps* of an élite formation. These fighting qualities were succinctly summed-up on 5 June as 4th Brigade moved up to the front through roads clogged with refugees and demoralized French units heading fast in the opposite direction. When a French officer suggested to Marine Captain Lloyd W. Williams that his men should join the retreat, the American replied, "Retreat, hell! We just got here!"

The Germans had exhausted themselves with their triple offensive, and the Allies now possessed the advantage of 300,000 American troops arriving every month. In July—August 1918, the Allied Supreme Commander, Marshal Foch, launched a limited series of counter-attacks — the British in the north, the French in the center and the Americans at the southern end of the line near Verdun — with the aim of squeezing the Germans out of the three huge salients they had driven into the Allied front in the spring and early summer.

On 18 July, Marines spearheaded the counter-offensive against the rail center at Soissons on the northern shoulder of the Marne salient. Casualties were heavy. On the first day of fighting, 2nd Battalion, 6th Marines, lost 50 percent of its men in 30 minutes. Overall casualties in pinching out the salient — a maneuver much favored by World War One commanders and much disliked by their men — were 1,972. Shortly after the Marines established themselves in positions overlooking the Soissons–Château-Thierry road, Marine General John A. Lejeune took command of 2nd Division, becoming the first Marine officer to command an Army division in combat.

In September the focus shifted to the Saint Mihiel salient, jutting across the Meuse between Verdun and Nancy, a thorn in the Allied side and a constant threat to movement in the Champagne area since 1914. The removal of this ugly feature was to be the preliminary to the final offensive against the Hindenburg Line — the immensely strong line of defenses built along an arc running through Lens–Noyon–Rheims — and the first battle of the war to be

fought independently by US 1st Army.

On 12 September seven American divisions — each more than twice the size of their French and German equivalents — attacked the two sides of the salient while the French nibbled away at its nose. Marine 4th Brigade was held back as a reserve. The attack caught the Germans in the middle of a tactical withdrawal, and I Corps, under General Liggett, USA, achieved its second-day objectives within a matter of hours. But the opportunity was not fully exploited and all but 4,000 of the 50,000 Germans threatened with encirclement slipped through the closing jaws of I and IV Corps. The Marines were then given the task of reducing the fortifications which screened the Hindenburg Line, and the salient was finally cleared by 16 September. Marine casualties were light, with 132 killed and 574 wounded. Some of the sharpest fighting was seen in the Bois de la Montagne by 2nd Battalion, 6th

Above: An artist's impression of Marines storming German machine-gun posts at the height of the battle for Belleau Wood.
Opposite: A wounded Marine is given first aid treatment before being stretchered to the rear.

Marines, who overran positions held by a substantial enemy force and then repelled four counter-attacks supported by artillery and gas.

The Germans had now abandoned the three salients they had carved out since March, but nowhere was their line broken. What had snapped was their morale and their faith in victory. The Allies launched a renewed counter-offensive on 26 September. The vital stroke was to be delivered by the Americans in the Argonne, with the aim of destroying the Germans' lateral line of communications running behind the Hindenburg Line.

After St Mihiel, 2nd Division had joined the French Fourth Army. Initially, considerable pressure had been exerted on General Lejeune to allow the Division to be broken up to provide shock troops for a number of French units. Lejeune succeeded in preserving the Division's integrity, and it was tasked with seizing Blanc Mont Ridge, on the heights of the Meuse, a primary position in the German line against which the French had been vainly battering for several weeks. The assault on Blanc Mont began on 3 October and, in two days of exceptionally hard fighting, 5th and 6th Marines, in a co-ordinated assault with an Army brigade of 2nd Division, overran the enemy positions. This action won the Marines their third citation from the French Army, earning them the right to carry the Croix de Guerre on their colors. The price of this honor was 2,538 casualties; in one company only 23 men remained of the 250 who had begun the battle.

The Marines ended the war serving with US 1st Army in the northern sector of the Argonne forest, punching their way through the Hindenburg Line. On 1 November, 4th Brigade took the last major strongpoint in the Line — "Freya Stellung" — after a lightning advance of more than five miles. By

Right: A signpost for the future — a Marine seaplane stationed in postwar Cuba takes off on an aerial survey mission.

9 November the Germans had withdrawn across the Meuse, and 4th Brigade was readying itself to attack across the river at Mouzon and Villemontry. On the 10th they crossed the Meuse under heavy enemy fire. They were preparing to renew their advance on the 11th — the day of the Armistice — when the guns fell silent.

Marine Aviation

The first Marine airman had been Lieutenant Alfred A. Cunningham, who in August 1912 became Naval Aviator No 5. In 1914, military aviation was still in the embryonic stage, but by April 1917 — when the United States entered the war — tactical air power over the battlefield, and strategic bombing operations against industry and centers of population, had become significant secondary factors in the waging of war.

However, when the United States declared war, the Marines' aviation section had only six officers (designated Naval Aviators), one warrant officer and 34 enlisted men. In September 1917 the first Marine Aeronautic Company was formed, and in January 1918 its 12 officers and 133 enlisted men were transferred to the Azores, from where it flew seaplanes on anti-submarine missions for the rest of the war.

In the spring of 1918 the 1st Marine Aviation Force was organized, with its headquarters at the former Curtiss Flying Field in Florida, Miami. On 30 July 1918 it despatched three landplane squadrons to France, with a fourth following in October. Upon their arrival they became the Day Wing of the Northern Bombing Group, stationed in northern France, flying the British-supplied De Havilland DH4, the best single-engined bomber of the war. From October 1918, Marines flew DH4s in operations in the Dunkirk area against German submarines and their bases at Ostend, Zeebrugge and Bruges. By the Armistice the Marine Aviation Force numbered some 2,500 officers and men and fielded 340 aircraft. They claimed 12 victories over enemy 'planes and delivered 52,000lb of bombs in 57 raids. They also performed the first recorded aerial resupply operation when they dropped food to a beleaguered French regiment, cut off for several days in the front line. It was a modest beginning for Marine aviation, which was to come into its own in World War Two.

3:FOUNDATION FOR VICTORY:
The Interwar Years 1919-1941

With the arrival of peace, the Marine Corps suffered the wholesale cutbacks affecting all the armed services which had fought "the war to end all wars". By July 1920, its wartime strength of 75,000 had been reduced to 17,165 officers and men on active duty.

In that month, General John A. Lejeune was appointed Commandant, a post he held until 4 March 1929. It was a difficult time for the Corps which, in common with all the services, had to cope with the drastically reduced funding resulting from the mood of isolationism which gripped the United States in the 1920s. Nevertheless, under Lejeune's shrewd leadership the Corps initiated a number of programs — most notably in the field of amphibious warfare — which laid the foundations of victory in the Pacific in World War Two.

Nicaragua and New Methods

Initially, however, active service was principally confined to the familiar troublespots of Haiti, Cuba and Nicaragua. In 1925 Nicaragua was once more threatened with civil war, and by May 1927 the Marine presence there had been reinforced to brigade strength.

The need for air support in Nicaragua's difficult terrain was quickly recognized, and in February 1927 VO-1-M, under the command of Marine Major Ross E. "Rusty" Rowell, was despatched to Managua, the Nicaraguan capital. VO-1-M was subsequently joined by VO-4-M and the two units combined to form an Aircraft Squadron, under Rowell's command, flying their DH4s on reconnaissance patrols over the two opposing Nicaraguan factions while the representative of the US government, Henry L. Stimson, was negotiating an armistice. In June a general pacification agreement was reached, the liberal insurgents laid down their arms, and a process of troop withdrawals began.

However, the United States' overriding role in Nicaragua's destiny had produced a strong armed nationalist movement — Ejercito Defensor de la Soberania Nacional (EDSN) — committed to ending all foreign influence in Nicaraguan affairs. Under the charismatic leadership of Augusto Cesar

Left: Bringing the boys back home; Red Cross women welcome men of 2nd Marines as they parade in New York on their return from France in 1919.

Marines pick up their rations while on board ship.

Sandino, the EDSN refused to accept the accord and moved into the inaccessible mountains of the north-east to wage a guerrilla campaign against the Marines and the newly-formed Nicaraguan National Guard.

The Marines established a number of garrisons in the jungle in an attempt to contain Sandino in the highlands. They were defeated by the rebels' superior knowledge of the terrain and the hostility of the local population, but the campaign provided a number of young Marine officers — among them Merritt A. Edson, Evans F. Carlson, and Lewis B. "Chesty" Puller — with an introduction to the tactics and techniques of jungle warfare. They were to make their mark in the Pacific in World War Two.

The fighting in Nicaragua also saw the beginnings of modern air–ground co-operation. On the night of 15 July 1927, Sandino launched a surprise attack with 600 men against the small outpost of Ocotal, held by 62 Marines under Captain Hatfield and 48 ill-equipped and poorly trained National Guardsmen, commanded by a Marine NCO. Sandino's attack was beaten off but, when dawn broke, the town was completely cut off. The nearest friendly outpost was 125 miles away, which meant 14 days of hard marching for a small Marine relief force. Sandino settled down to wait for the defenders of Ocotal to run out of ammunition.

Two aircraft of Rowell's unit — now known as VO-7-M — made aerial contact at 1010 hrs on 17 July, strafing Sandino's guerrillas before returning to make a report. Rowell had five DH4s ready for operations and these immediately took off, armed with 17-lb fragmentation bombs.

Rowell was one of the postwar pioneers of dive-bombing and all his pilots were thoroughly trained in this technique.

The time had come to put Rowell's dive-bombing theories to the test. The DH4s reached Ocobal against the dramatic backdrop of a threatening tropical storm, circling the target area once at 1,500ft and then forming their bombing column. Rowell led the attack, pulling out of his dive at 600ft. Sandino's men had never been exposed to serious aerial attack and disdained to take cover, losing as many as 80 men in the first attack. Subsequent dives were made from 1,000ft with the DH4s pulling out at 300ft. Sandino's men broke and fled. Although dive-bombing by individual aircraft had been a feature of the closing months of World War One, this was the first use of co-ordinated dive-bombing techniques in combat.

The value of air–ground support was underlined in January 1928, when Marines besieged at Quilali were sustained from the air by First-Lieutenant (later General) Christian F. Schilt. Using an improvized airstrip cut through the center of the town, Schilt flew in supplies and ammunition and flew out the wounded, his ten missions winning him a Medal of Honor. As the campaign against Sandino wore on, it became standard procedure to use aircraft in support of patrols and garrisons.

Sandino's army fought on for six years, and by 1933 the war had spread to 10 of Nicaragua's 16 provinces. Unable to secure a decisive victory against the rebels, and under considerable domestic pressure to end American involvement, the Marines were withdrawn from Nicaragua in 1933. Sandino was assassinated a year later, but his ghost — in the form of Nicaragua's Sandinista regime — came back to haunt the US government in the 1980s.

The Orange Plan

In the interwar years there was a ferment of new ideas about the tactics and technology of future conflicts. In Germany the revolutionary theories of armored warfare advanced by the British military writers Fuller and Liddell Hart were taken up by a brilliant young Army officer, Heinz Guderian, principal architect of Blitzkrieg warfare. The Italian General Douhet and the merican polemicist General William Mitchell argued that air power, in the form of massive fleets of strategic bombers, would become the single, dominant factor in war, rendering all others obsolete. Mitchell's enthusiasm for the long-range bomber was eventually proved correct by the global application of his theories, and confirmed his tireless advocacy of the importance of Alaska as the strategic key to the Pacific.

Another less-famous military visionary obsessed with the perils and potential of American strategy in the Pacific was a Marine Corps staff officer, Major (later Lieutenant-Colonel) Earl H. Ellis. As early as 1913 Ellis had delivered a series of lectures at the Naval War College in which he warned of a future war with Japan, the outcome of which would hinge on the possession of the islands and atolls of the trans-Pacific. To meet this threat, Ellis evolved a wide-ranging and revolutionary approach to the waging of amphibious warfare.

Unlike Mitchell, Ellis was not a voice in the wilderness. Prior to World War One, War plans derived their titles from the codenames of the probable enemy. Japan was codenamed Orange, and the strategic concept and missions to be followed in the event of a war with that country were contained in the Orange Plan. The Orange Plan was constantly revised as the international situation changed.

Before 1914, the broad strategy was for the Army to defend the Philippines until the Fleet could carry reinforcements across the Pacific. Naval strategists were keenly aware that Japan would almost certainly seize Guam before a relief force could be despatched, depriving the US Navy of its only Fleet anchorage between Pearl Harbor and Manila Bay. Thus Guam would have to be retaken or another island occupied as a coaling and repair station.

After World War One, Orange assumed new dimensions when Japan gained control of the former German possessions in the Marshalls, Carolines and Marianas. The Philippines were now more vulnerable, and Guam ringed with Japanese outposts. The Orange Plan came under review.

In 1920 the Office of Naval Intelligence proposed a study of the possibility of a trans-Pacific war against Japan, and various agencies were directed to assist the study with plans of their own. In 1921 Ellis was assigned to the newly-formed Division of Operations and Training at Headquarters, Marine Corps. It was in this year that Ellis' 1913 paper, "Naval Bases: Location, Resources, Denial of Bases, Security of Advanced Bases", was published, along with his "Advanced Base Operations in Micronesia in 1921". The latter became the basis of Operation Plan 712D, the Marine Corps' contribution to the Orange Plan, approved by the Commandant, General John A. Lejeune, in July 1921.

Ellis' plan concentrated on one aspect of a future war against Japan — the seizing of a base which would be urgently needed by the Navy as a coaling or a repair station. Focusing his attention on the islands in the Marshalls Group, Ellis outlined the tactics to be used in such an operation. He wrote: "It will be necessary for us to project our fleet and landing forces across the Pacific and wage war in Japanese waters. To effect this requires that we have sufficient bases to support the Fleet, both during its projection and afterwards. To effect [an amphibious landing] in the face of enemy resistance requires careful training and preparation, to say the least, and this along Marine lines. It is not enough that the troops be skilled infantrymen or artillerymen of high morale; they must be skilled watermen and junglemen who know it can be done — Marines with Marine training."

The practical application of Ellis' theories was limited by the rudimentary nature of the equipment then available to the Corps, but he made a number of significant recommendations, not the least of which was the stress he placed on naval gunfire support for troops fighting ashore. Crucially, Ellis' approach radically altered thinking about the role which the Marines were expected to play. Ellis outlined a scenario in which they would no longer be used primarily to defend advanced bases; now they would seize bases from the enemy.

This shift in the Marines' priorities was underlined in February 1921 when Lejeune suggested that the Corps' primary wartime function was to act as a mobile force accompanying the Fleet for operations in support of the Fleet. Addressing the Naval War College in 1923, Lejeune laid great stress on the importance of the Marine Corps as a force in readiness: "On both flanks of a Fleet crossing the Pacific are numerous islands suitable for utilization by an enemy for radio stations, aviation, submarine or destroyer bases. All should be mopped up as progress is made . . . The presence of an expeditionary force with the Fleet would add greatly to the striking power of the C-in-C of the Fleet . . . The maintenance, equipping and training of its expeditionary force so that it will be in instant readiness to support the Fleet in the event of war, I deem to be the most important Marine Corps duty in time of peace."

Ellis did not live to see his ideas take concrete form. He died in mysterious circumstances in 1923 while making a lone reconnaissance of the Japanese-held islands in the Pacific, having almost certainly confirmed that they were being fortified in contravention of the 1921 Five Powers Naval Treaty. It was left to others to develop his technical and tactical guidelines into a coherent body of doctrine.

In the immediate postwar years, precedent was not encouraging. Waterborne landings in World War One had been uniformly chaotic, with the Gallipoli landings of 1915 providing a bloody reminder of what could happen when a large-scale amphibious operation unravelled in the face of determined opposition. The majority of strategists remained deeply sceptical about the usefulness of amphibious operations. However, after an internal study of the Gallipoli campaign, the Marine Corps concluded that it was sheer incompetence, rather than any inherent flaw in the amphibious idea, which had undermined the operation.

Amphibious exercises conducted in Cuba between 1922 and 1925 confirmed the inadequacy of conventional tactical doctrine and the converted wooden Navy launches which were used as landing craft. But the Corps' expeditionary duties during the rest of the 1920s diverted its attention from

Above: A Marine mans a heavy machine-gun aboard a river patrol boat during the postwar occupation of the Rhineland.

amphibious warfare, and it was not until the following decade that the momentum was regained.

As the Marine Corps withdrew from its policing operations in Central America and the Caribbean, more troops became available for training in landing operations. However, before any real progress could be made, a substantial and permanent force of Marines — with its own command staff — would have to be organized for the purpose. The Assistant Commandant, General John H. Russell, proposed that the old "Expeditionary Force" should be replaced with a new formation called either the "Fleet Marine Force" or the "Fleet Base Defense Force". The new force would form an integral part of the US Fleet, under the operational control of the Fleet Commander when embarked on vessels of the Fleet or engaged in Fleet exercises ashore or afloat. These proposals were rapidly approved by the Commandant and the Chief of Naval Operations, and the designation Fleet Marine Force (FMF) chosen. The decision became official with the issuing of Navy Department General Order 241 on 8 December 1933.

The creation of the FMF provided the Marine Corps with the tactical framework within which it could execute the primary war mission of seizing and securing advanced bases. In 1934 the Corps produced a textbook whose purpose was to embody the theory and practice of landing operations, *The Tentative Manual for Landing Operations*. It was approved by the Chief of Naval Operations for "temporary use . . . as a guide for forces of the Navy and the Marine Corps conducting a landing against opposition". In spite of its engagingly modest title, this manual laid the foundations for all modern amphibious thinking in the American armed forces. In 1938 the Navy adopted it as official doctrine with the issuing of Fleet Training Publication 167; and the Army followed suit in 1941 with Field Manual 31-5. The fundamental elements of amphibious warfare covered by *The Tentative Manual for Landing Operations* included: command relationships; naval gunfire support; close air support; ship-to-shore movement; combat unit landing; and the organization of shore party personnel.

Tools for the Job

Flowing directly from the evolution of a body of amphibious tactics was the development of the technology to put it into practice. The overriding requirement of all amphibious operations is to get the troops on to the beaches with as few casualties as possible. By the early 1930s, pioneering work on amphibious craft was underway in the United States.

In 1933 a retired manufacturer, Donald Roebling, developed a remarkable amphibious vehicle for rescue work in the swamplands of the Florida Everglades. His design comprised an aluminum hull fitted with tracks which, by paddle-wheel action, propelled the amphibian through the water and enabled it to move across country. In 1937 Roebling's "Alligator" came to the attention of the Marine Corps, whose Equipment Board authorized a series of tests. In 1939 the board ordered three "Alligators" and put them through their paces at Quantico Marine base. Encouraged by the results and the basic soundness of the design, the Corps made some minor modifications to the amphibian and then gave Roebling a contract to produce the vehicle in quantity. This marked the birth of the Landing Vehicle, Tracked (LVT).

Subsequently the Navy Bureau of Ships produced an LVT design of its own, passing the blueprints to the Borg-Warner Corporation for manufacture. The principal modifications were the adoption of a sprung suspension and the installation of the engine and drive train from the M3 tank. Designated LVT2, the vehicle went into production in 1942.

At the same time, the executive officer of the Marine Equipment Board, Major Ernest E. Linsert, was developing an armored amphibian, the Landing Vehicle, Tracked (Armored), or LVT(A), based on the original "Alligator" design. Turrets made of steel casings housed the LVT(A)'s main armament: a 37-mm gun and 7.62-mm machine-gun mounted in an M3 turret; one 12.7-mm gun in each of its two side turrets; and two fixed 12.7-mm guns fired by the driver manipulating buttons on the steering levers. The first LVT(A)s came off the production line in the winter of 1940/41.

Although the tracked amphibians were destined to play a vital role in the Pacific War, the rapid build-up of men, machines and materiel could not have been achieved without the development of a new type of landing craft. The pioneer in this field was Andrew J. Higgins, who in the mid-1920s produced the "Eureka" boat for fur trappers

Above: The Douglas SBD Dauntless dive-bomber, all of the first batches of which went to the Marine Corps. Although obsolescent by the Battle of Midway (June 1942), the rugged Dauntless was a war-winner, sinking more Japanese shipping during the Pacific war than any other Allied type. The SBD-1 had a maximum speed of 252mph, ceiling of 24,300ft and range (as dive-bomber) of 456 miles and as a scout bomber of 773 miles. Armament comprised two 0.3-in Browning machine-guns fixed above the nose; one (later two) 0.3-in Browning manually aimed from the rear cockpit; and one bomb or other store of up to 1,000lb on a swinging crutch under the belly. Additionally, there were outer-wing racks for two 100-lb bombs or depth charges.

in the Louisiana swamplands. To cope with the region's shallow, weed-infested waters, the boat had shallow draft and a protected propeller.

In 1934 Higgins took his "Eureka" design to the Marine Corps, who were interested but had no money to fund a long-term development project. Undaunted, Higgins continued to liaise with the Corps and by 1939 had produced an improved landing craft which in a series of tests consistently outperformed all comparable vessels. It retained one serious design drawback — the troops had to disembark by jumping over the side. In April 1941 Higgins visited the Marine base at Quantico, where he agreed to redesign the craft with a ramp bow. Working with Brigadier-General Moses, the head of the Marine Equipment Board, Higgins produced a design which met the Corps' specifications. At the beginning of 1942 the Board of Ships authorized full-scale production of the Higgins Boat.

Marine Aviation in the 1930s

The 1930s also saw significant changes in Marine aviation. Flying aircraft off ships was one such change, and from 1931 to 1934 VS-14M and VS-15M, the first Marine squadrons to become part of the Fleet air organization, saw service aboard the carriers *Saratoga* and *Lexington*.

Marine aviation also adapted to the establishment of the FMF and the emphasis now placed on the seizure of advanced bases in the event of war. In 1935 the aviation section at Headquarters, Marine Corps, was removed from the Division of Operations and Training and established as an independent section under the Commandant. In 1936 it became a Division under a Director of Aviation, who served as an advisor to the Commandant on all aviation matters. The Director also acted as liaison officer between the Marine Corps and the Navy Bureau of Aeronautics.

The Marine Corps remained completely dependent on the Navy for its aircraft and equipment. But it was the Marines who in June 1940 took delivery of the first 57 Douglas SBD-1 Dauntless dive-bombers. The Dauntless was to be one of the most important US combat aircraft of World War Two, sinking more Japanese shipping than any other Allied weapon, stopping the Imperial Fleet at Midway in June 1942 and later playing a major role in the Solomons campaign.

By the summer of 1939 there were 210 officers and 1,142 enlisted men on active duty with the Marine Corps' aviation arm. In June 1940, Congress authorized the Navy's 10,000-'plane program, out of which the Corps was allotted 1,167 aircraft. This led to the preparation of plans for four groups, each comprising 11 squadrons. Landing exercises in 1941 indicated that a single division making an amphibious landing would require air support from a minimum of 12 fighter, eight dive-bomber, two observation and four utility squadrons. However, progress towards achieving this level of support was slow. When the two brigades of Marines were reorganized into divisions in February 1941, their air support groups were redesignated 1st and 2nd Marine Aircraft Wings (MAWs). When war broke out, there was still only one group in each wing and a total of 251 aircraft and 708 pilots in Marine aviation.

On the eve of war, the Marine Corps was spread far and wide across the globe. Some 18,000 of its 65,000 officers and men were deployed overseas. A brigade built up around 6th Marines (reinforced) had been despatched to Iceland to relieve a British garrison and pave the way for the arrival of more US troops. There were Marines on duty in China, the Philippines, Guam, Wake Island, Midway, American Samoa, Panama and a number of British islands in the Caribbean. Another 4,000 Marines were serving aboard Navy ships or at Navy stations, and 20,000 were in training at land bases. And there were some 4,500 Marines on the Hawaiian island of Oahu, on duty at the great US naval base — Pearl Harbor.

Boeing F4B biplanes of VF-10, part of the West Coast Expeditionary Force, c. 1930.

4:THEIR FINEST HOUR:
World War Two

JAPAN IN THE ASCENDANT: WORLD WAR TWO

Just before dawn on 7 December 1941, the first wave of Japanese aircraft — 43 Mitsubishi A6MZ Zero-Sen fighters, 89 Nakajima B5NA "Kate" level-altitude bombers and 81 Aichi D3A2 "Val" dive-bombers — took off from the carriers of Admiral Nagumo's task force, lying some 250 miles north of the US Pacific Fleet's base at Pearl Harbor. The Japanese aircraft were picked up by radar as they approached Pearl Harbor, but no action was taken. In the calm, clear light of early morning, the naval base remained to all intents and purposes undefended: battleships berthed alongside each other in the harbor; nearly all the anti-aircraft guns unmanned and the ammunition under lock and key. The attack on Oahu began at 0755 a.m. and continued, with a brief lull, for nearly two hours. Using torpedoes and specially-adapted armor-piercing bombs, the Japanese inflicted most damage in the first five minutes of the attack, when resistance was non-existent and the targets were not obscured by smoke. Four battleships were sunk and a fifth, *Nevada*, crippled and beached. Three destroyers and a minelayer were also sunk and 188 aircraft destroyed, including all but one of the 48 'planes of Marine Air Group 21 (MAG-21). There were 2,403 American dead, including 112 Marines. Japanese losses were 29 aircraft and 555 men.

Superficially, the pre-emptive strike on Pearl Harbor had been an overwhelming victory. Hindsight enables us to see that on 7 December the Japanese took their first step on the road to defeat. The greater part of the US Navy escaped unscathed. After Pearl Harbor the US Navy still retained 18 heavy cruisers — 12 of them in the Pacific — 168 destroyers and, crucially, 112 submarines. Above all, the three aircraft carriers of the Pacific Fleet — *Enterprise*, *Saratoga* and *Lexington* — were absent from Pearl Harbor when the Japanese attacked. The urgent need for Marine aircraft to strengthen outpost defenses had sent *Lexington* and *Enterprise* to sea on aircraft-ferrying missions. On 7 December *Lexington* was en route to Midway with aircraft of Marine

Left: The face of the Pacific war. Bitter fighting in the Marshall Islands, February 1944.

Scout Bomber Squadron 231 (VMSB-231). She turned back when news of the attack was received. On that fateful day *Enterprise* was steaming back to Pearl Harbor after flying off the fighters of VMF-211 to reinforce the defenses of Wake Island. She should have tied up in Pearl Harbor at about 0730 hrs on 7 December, but she was delayed by refueling problems and was some 200 miles west of Oahu when the attack went in. Warned of the situation by some of her aircraft, which had encountered the first Japanese wave, the carrier sailed westwards and out of danger.

Guam and Wake

In the Japanese war plans, the small concentration of forces based on Truk was earmarked to secure the islands of the south-west Pacific. The first phase of operations including the seizing of the American islands of Guam and Wake.

As US strategists in the 1920s had anticipated, Guam was virtually indefensible. The most southerly of the major islands in the Marianas, Guam was hemmed in on all sides by Japanese possessions, particularly to the south in the Carolines. To defend Guam's 225 square miles there was a garrison of 153 Marines, armed with nothing heavier than .30-caliber machine-guns and reinforced by about 80 local militiamen. Within hours of the attack on Pearl Harbor, Guam was subjected to an air offensive which lasted two days before the Japanese put 6,000 troops ashore on the morning of 10 December. The tiny garrison was overwhelmed within an hour, the Marines surrendering to prevent the possibility of Japanese reprisals against the local population. Four Marines and 15 Guamanians had died in the defense of the island.

Wake Island was a different proposition. Wake is a small bleak atoll of three islands and a lagoon, 2,000 miles from Hawaii and over 600 miles from the nearest Japanese base. Before the war it had been a Pan-American Airways staging post. Any major or prolonged Japanese attempt to seize Wake would run the risk of a fleet engagement. The likelihood of such a clash depended on timely intelligence and the ability of the Wake garrison to hold out against the Japanese attack for a length of time.

However, Wake was only marginally more heavily defended than Guam. In October 1941, command of the Marine garrison of 178 officers and men was assumed by Colonel James Devereux. On 2 November he was reinforced by 200 Marines led by Major George H. Potter, and at the end of the month Commander Winfield Scott Cunningham of the US Navy arrived at Wake to take overall command of the island.

Devereux remained in command of the

Left: The battleship *Pennsylvania* and the destroyers *Cassin* and *Downes* in the aftermath of the attack on Pearl Harbor. *Pennsylvania* was in dry dock at the time of the attack and thus saved from the attentions of Japanese torpedo-bombers. *Cassin* and *Downes* took the brunt of the enemy dive-bombing attacks. *Pennysylvania* lived to fight another day and played an important part in The Battle of Leyte Gulf, in October 1944, which saw the last battleship-against-battleship engagement.

MARINE DEFENSE BATTALIONS

In the years following the formation of the Fleet Marine Force (FMF), the prevailing mood of isolationism in the United States restricted the funding of the armed services by Congress. With some ingenuity, the Commandant of the Marine Corps, Major-General Thomas Holcomb, argued that an increase in the Corps' manpower would bolster the nation's defensive, rather than offensive, capability. Thus the Defense Battalion Program eventually got underway on 10 October 1939 with the activation of 3rd Defense Battalion. The aim was to provide the Corps with a force capable of seizing and securing bases for the US Fleet. Tables for four different types of Defense Battalion appeared in October 1941. Each had a Headquarters and Service Battery, and a 90-mm or 3-in anti-aircraft artillery (AAA) group. The addition of two of the following components would complete the battalion's organization: a 155-mm artillery group; a special weapons group; a machine-gun group; or a 7-in artillery group. Combat strength of the battalion was approximately 1,000 officers and men, including a naval component of 25 doctors, dentists and hospital corpsmen. When the United States entered the war, a number of understrength Defense Battalions were stationed in the Pacific, notably at Wake and Midway. In the 1944 drive to increase the strength of the six Marine divisions, 17 of the 20 Defense Battalions were redesignated anti-aircraft battalions. Where possible, their men were transferred to AAA units; the remainder were retrained as field artillerymen and transferred to line artillery regiments.

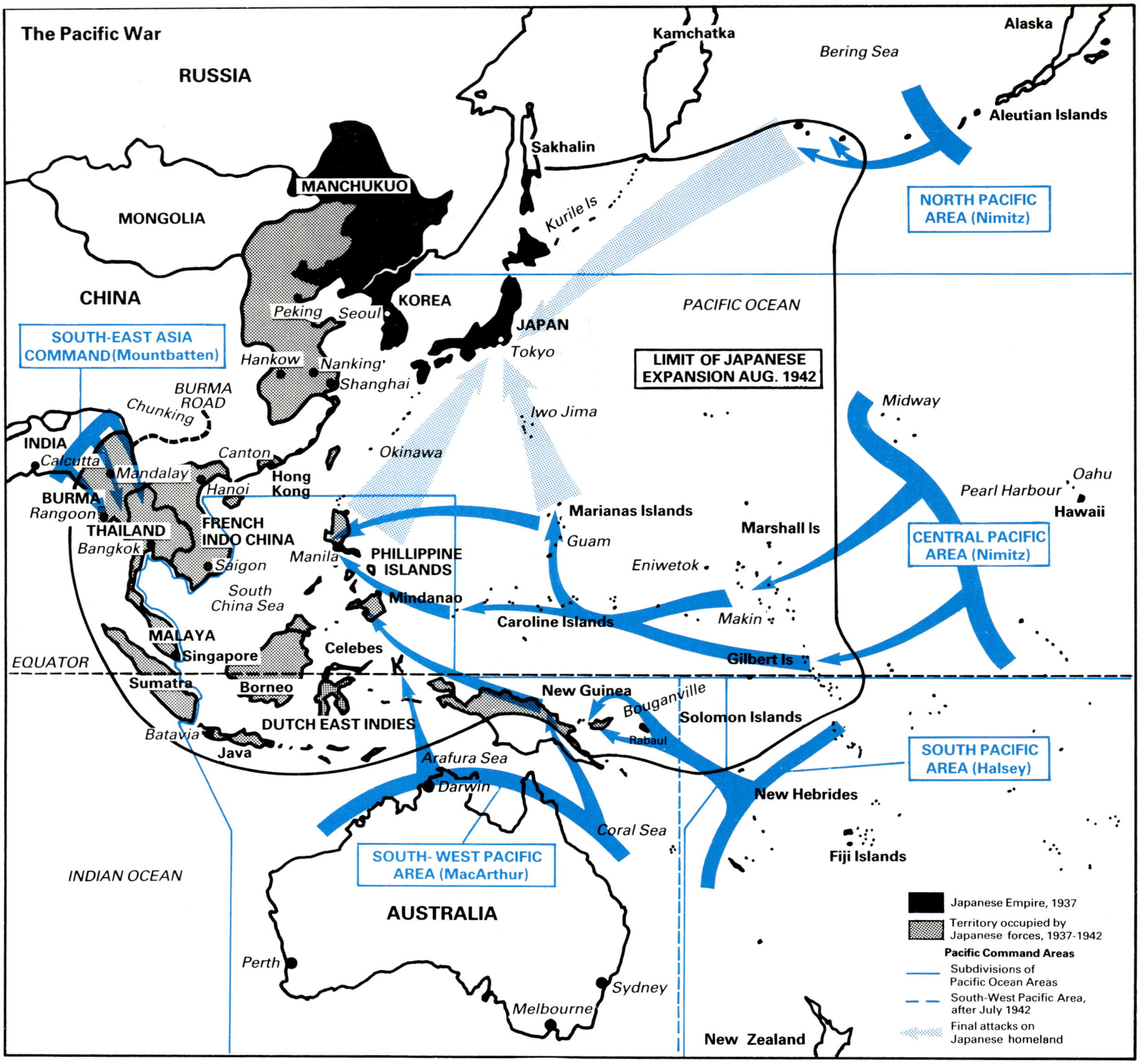

The map shows the strategic situation in the Pacific in February 1944, with the taking of the Gilbert and Marshall Islands, but before the trap closed completely on Rabaul.

Marine detachment, and was responsible for co-ordinating the operations of the Defense Battalion and the Marine Fighter Squadron VMF-211. The latter, led by Major Paul A. Putnam and comprising 12 Grumman F4F Wildcats, had flown in from the flight deck of the carrier *Enterprise* on 4 December.

The Marines' artillery consisted of six 5-in coastal guns and 12 3-in anti-aircraft pieces. But the understrength Defense battalion could not man these guns simultaneously and possessed only half the numbers sufficient to operate the 0.5-in and 0.3-in machine-guns which were to be used for air and ground defense respectively. An SCR-270B search radar had been promised but not delivered. Wake's improvised air-raid warning system was supplied by Marines stationed atop two 50ft water towers.

The air raids were not long in coming. At noon on 8 December, 30 Japanese bombers appeared over Wake, dropping a dense pattern of bombs on the atoll's airfield. Seven Wildcats were destroyed on the ground, and the squadron's 25,000-gallon fuel tank erupted in a ball of flame. Twenty-three Marines died in the attack and 11 were wounded. Crossing the lagoon, the bombers strafed the Pan American facilities, setting fuel tanks ablaze and killing 10 American civilians.

The bombers reappeared at 1145 on the

9th. Twenty-seven aircraft flew in from the south and were met by four patrolling Wildcats, which claimed one kill. Fourteen of the bombers flew away with heavy flak damage. On the 10th, 18 bombers arrived over Wake, two of which were shot down by Captain Henry Elrod.

Meanwhile, a Japanese assault force was closing on the atoll. The preparations for the attack were uncharacteristically slipshod, influenced perhaps by exaggerated reports of the success of the bombing raids and a fatal underestimation of the resolve of the Marine garrison. Very few troops — less than 450 in the assault phase — were embarked for the landings at dawn on the 11th. The Japanese covering force comprised three lightcruisers (the flagship *Yubari*, *Tatsuta* and *Tenryu*) and six destroyers (*Hayate*, *Kisaragi*, *Mochizuki*, *Mutsuki*, *Oite* and *Yayoi*).

Although most of the garrison's Wildcats had been destroyed, the 5-in and 3-in guns remained intact. As the Japanese assault force approached Wake, the Marines held their fire until the range had closed to 4,500yds, at which point Colonel Devereux gave the order to commence firing. The 5-in guns on Wilkes Island and Peacock Point, commanded by Lieutenant Clarence A. Barninger, scored two hits amidships on the port side of the cruiser *Yubari*. These were followed by two more shells which slammed into the ship slightly aft of the first two hits. *Yubari* limped away billowing smoke and steam. One of the escorting destroyers attempted to cover her withdrawal by laying down a smokescreen, but was herself driven off when a 5-in shell exploded on her forecastle.

One of the transports, a converted destroyer, was hit and crippled and another transport damaged. In an effort to draw the Marines' fire from the transports, three of the destroyers immediately closed to engage the shore batteries. Battery L, commanded by Lieutenant John A. McAlister, quickly found the range and blew *Hayate* out of the water. She was the first Japanese warship to be sunk by US forces in World War Two.

Oite and *Yayoi* sustained hits during these exchanges, also coming under attack from Wake's four Wildcats. Able to carry only two 100lb bombs, the fighters had to return to the airfield to re-arm after each bombing run. At 0730hrs, during his final sortie, Captain Elrod scored a direct hit on *Kisaragi*, setting off her depth charges. She sank almost immediately with all hands. The Japanese withdrew. Without being able to put a man ashore, they had lost two ships and 700 men. Amidst the tide of victory in the Far East, this had been a uniquely disastrous operation.

Inevitably, the Japanese redoubled their efforts to seize Wake, whose capture was vital to complete their defensive perimeter across the width of the Pacific and to prevent the atoll being used by the Americans as a jumping-off point for future operations in the Western Pacific. Moreover, the possession of Midway would facilitate future Japanese operations against Midway and the western Hawaiian group. The failure of the attack on 11 December had made a second operation against Wake markedly more hazardous, as it gave the Americans a chance to reinforce the garrison and deploy a powerful task force to counter the Japanese assault.

This the Americans failed to do. The ill-fated Admiral Kimmel, C-in-C of the Pacific Fleet at the time of Pearl Harbor, had ordered the reinforcement of Wake, instructing *Saratoga* to support the atoll's garrison and ordering *Lexington* and *Yorktown* to stand by their sister carrier. But *Saratoga* was delayed in Pearl Harbor by refueling problems, and then lost two days topping up her destroyers at a time when speed was of the essence. On 17 December, Kimmel was removed from his command and temporarily succeeded by his deputy, Admiral William S. Pye, pending the arrival of the new C-in-C Pacific Fleet, Admiral Chester W. Nimitz. Unsure of the strength of Japanese forces in the area of Wake, Pye shrank from risking his carriers so soon after the débacle at Pearl Harbor. Ultimately, Wake was expendable, but the carriers were not. Pye decided against the relief of Wake.

Meanwhile, the atoll was being pounded twice a day by Japanese aircraft: at noon by land-based bombers and at dawn by flying boats. On 14 December the tail of one of the two remaining Wildcats was set on fire during a raid. Three men from the squadron's engineering section, Second Lieutenant John Kinney, Technical Sergeant William Hamilton and Aviation Machinist's First Mate James Hesson, left their cover, removed the undamaged engine from the burning aircraft and dragged it clear. By ingeniously cannibalizing the squadron's damaged 'planes, they had within three days provided VMF-211 with four operational Wildcats.

The damaged destroyer *Downes* in the dry dock at Pearl Harbor. She was repaired and saw service in the Pacific.

On 20 December, Wake's battery positions were hit by 29 Japanese Navy bombers escorted by 18 Zero fighters. This was the first time that the atoll had been attacked by carrier-based aircraft and a sure indication that an enemy task force was closing on Wake. The assault force consisted of the survivors of the 11 December defeat, supported by four heavy cruisers from Guam, the carriers *Hiryu* and *Soryu* from Admiral Nagumo's Strike Force and the seaplane tender *Chitose*. It was a powerful group but, with grim irony, inferior to US naval forces in the general area.

By the evening of 22 December, Wake's remaining Wildcats had been destroyed. At 0120 hrs on the 23rd, 1,000 Japanese Special Naval Landing Force troops came ashore at four points on the southern shore of Wake, skilfully seeking the security of dead ground for their landfall. Prudently, the heavy cruisers stayed out of range of the shore batteries. As the Japanese moved up the beach, a 3-in shell from one of the Marine batteries blew up an ammunition magazine on one of the destroyer-transports, turning the night into day.

The Marines put up a ferocious resistance, and there was bitter hand-to-hand fighting on the water's edge. The dawn came up, revealing 17 enemy warships ringing the atoll. Three destroyers steamed in to finish off the gallant shore batteries, and in a final act of defiance the 5-in guns hit and heavily damaged one of them, *Mutsuki*. By 0700 hrs all the Marines' forward positions had been overrun, and Commander Cunningham ordered them to lay down their arms. The order went out: "Cease firing. Destroy all weapons. The island is being surrendered." After performing this melancholy task the Marines sat down in their positions and — perhaps mindful of what lay ahead — began to eat as much as they could.

The garrison had held out for 15 days after the first raid. During the raids, shelling and assault on the island, they had suffered 81 killed and wounded and had inflicted approximately 1,000 casualties on the enemy. Four Japanese ships had been sunk and eight more damaged. Twenty-one enemy aircraft had been shot down by the island's fighters and anti-aircraft guns. On 12 January the Marines were loaded aboard the *Nitta Maru* and sailed off into four long years of captivity. Their heroic example prompted President Roosevelt to declare that millions had "been inspired to render their own full share of service and sacrifice".

The Fall of the Philippines

The opening weeks of the Pacific war were characterized by the brilliant Japanese concentration and co-ordination of all arms against the enemy's points of maximum weakness. On 8 December, heavy bombing raids crippled US air power in the Philippines. By the end of the afternoon, for the loss of just seven Zero fighters, the Japanese had destroyed 103 US aircraft, most of them on the ground and over half of them fighters. When General Homma's XIV Army landed on the northern tip of the island of Luzon on 10 December, American aircraft had ceased to pose a serious threat. Two days later a Japanese force from Palau captured Legaspi, in the extreme south of Luzon. The main landings went in on 22 December in the Lingayen Gulf. After establishing a strong beach-head, the Japanese advanced south, meeting only limited pockets of stiff resistance.

Without adequate naval and air support, the policy of forward defense against the Japanese adopted by General Douglas MacArthur, commanding US Army forces in the Far East, proved a forlorn hope. On 23 December, MacArthur withdrew from Manila into the Bataan peninsula. The Japanese entered the Philippines capital on 2 January and then confronted MacArthur at the northern end of the Bataan peninsula. MacArthur remained confident, claiming, "The Japanese may have got the bottle but we have the cork."

Marines of the 4th Division, recently arrived from Shanghai, were not placed in the immediate front-line. They were assigned the task of defending the island fortress of Corregidor at the entrance to Manila Bay, and two of its three supporting forts, Hughes and Drum. It was not long before these "impregnable" bastions became the last-ditch defenses of the Philippines.

In the Bataan peninsula, the Japanese threw in their first attack on 9 January. They were beaten back with heavy losses on both sides, but by the end of the month the Americans had been forced back to their reserve line, running across the peninsula from Bagac to Orion. More than once in the last week of January the Japanese attempted amphibious landings against the south-west coast of Bataan, but two Japanese battalions were destroyed on their beach-heads. Two months of stalemate followed, during which both sides began to feel the effects of disease and malnutrition.

On 12 March, MacArthur reluctantly left his headquarters in Corregidor for Australia, handing over command to General Jonathan Wainwright. On 3 April the Japanese launched a new attack, breaking through the American line and turning the flank of 2nd Corps. On 9 April the 80,000 US and Filipino troops on Bataan surrendered. The Japanese Army had not prepared for a capitulation on this scale and adopted the most expedient course, marching the troops 65 miles to a railway line and then shipping them to Camp O'Donnell. En route, the Japanese treated their prisoners with extreme brutality, giving them no food or drink for five days, shooting stragglers and inflicting many other hardships. This was the infamous Bataan Death March, in which about 16,000 US and Filipino troops died and for which General Homma was tried and executed after the war.

After the surrender, about 2,000 troops escaped to Corregidor. Japanese aircraft had made their first raid on the "Gibraltar of the East" on 29 December. Three days later the garrison was placed on half rations. From the end of March, the Japanese began a regular artillery bombardment of the island which increased in intensity after the surrender on Bataan. For the men of 4th Marines, dug in behind Corregidor's beach fortifications in foxholes, it was "like living in the center of a bullseye". The bombardment continued day and night, churning up the defenses faster than they could be rebuilt and transforming the "Rock" into a landscape of rubble. It reached a crescendo on 29 April, when more than 150 heavy guns thundered out a murderous salute to the Emperor's official birthday, setting the remaining undergrowth ablaze, throwing up such a pall of dust that the mainland was completely obscured, and smashing the last vestiges of co-ordinated defensive zones on Corregidor's beaches.

Conditions on Corregidor were intolerable. The only secure refuge from the ceaseless bombardment was the Malinta Tunnel, a

A Marine unit in the Bataan peninsula, March 1942.

Right: Admiral Yamamoto, the strategic genius who promised, and delivered, to the Emperor Hirohito six months of victory between December 1941 and June 1942. He never fully recovered from the defeat at Midway and was shot down and killed by US fighters on 18 April 1943 while flying from Rabaul to Bougainville.

bomb-proof underground complex located at the base of Corregidor's long tadpole-like tail. But it was now fetid with the stench of hundreds of wounded, jammed into its stifling, dust-choked side chambers. Malnutrition, malaria and dysentery had remorselessly eroded morale. On 3 May General Wainwright radioed MacArthur that, with less than five days of water left, the situation was critical. On 4 May, 16,000 shells fell on Corregidor, and that night Homma despatched assault boats carrying 2,000 men against the fortress. Their landings were preceded by another furious barrage, but the assault craft were caught by a strong current and missed their landing points by more than a mile. When the Japanese came ashore, they were met with fierce enfilading fire from 1st Marine Battalion, which took full advantage of the rising moon to inflict heavy casualties.

The men of 4th Marines — their numbers swollen to nearly 4,000 officers and men by Army, Navy and Filipino troops — fought valiantly, sustaining nearly 700 casualties before their commander, Colonel Samuel L. Howard, ordered them to surrender. Their last pockets of resistance in the Malinta tunnel area were overcome on 6 May. The survivors joined the infamous Death March into captivity.

The Long Road Back

In a six-month explosion of military energy, the Japanese had overrun Malaya, taken Singapore, bundled the British out of Burma, swallowed up the Dutch empire in the East, isolated China, menaced the borders of India, brushed the north coast of Australia and gained control of nearly half the Pacific. They had acquired vast oil, tin and rubber resources and their safe passage to the Japanese home islands.

There was now an immense buffer zone against an American counter-attack, but its very size posed strategic problems which the Japanese failed to resolve. If the United States, with its immeasurably superior economic and industrial resources, could not be decisively defeated, Japan would be committed to the forward defense of a hugely-extended perimeter, where the enemy would enjoy the luxury of choosing the time and place of attack. Admiral Yamamoto, C-in-C of the Japanese 1st Combined Fleet and architect of Japan's stunning victories, had warned that he could promise only six months of victory. The six months were now up.

In May and June 1942, Japanese expansion was checked at two naval battles, Coral Sea and Midway. At Coral Sea (7/8 May) — the first of the carrier battles of the Pacific war — the Japanese were thwarted in their attempt to outflank New Guinea, seize Port Moresby and isolate Australia. Still wrestling with their self-imposed strategic dilemma, the Japanese then implemented an immensely elaborate plan — involving the movement of eight separate task forces — for the capture of Midway Island, which offered a base within striking distance of Hawaii. It was Yamamoto's aim to draw the US Pacific Fleet into a decisive naval battle, but the tables were about to be turned on the triumphant Japanese.

Yamamoto's plans had been fatally compromised by the Americans' breaking of the Japanese Fleet Code, which gave them ample foreknowledge of the attack on Midway on 4 June. The resulting battle was indeed decisive, but it was Admiral Nagumo's 1st Carrier Strike Force which was destroyed. On 4 June, the US Task Forces 16 and 17 sank three Japanese carriers — *Kaga*, *Akagi* and *Soryu* — for the loss of the carrier *Yorktown*. On the following day the crippled Japanese carrier *Hiryu* was scuttled.

The victory at Midway was won principally by carrier-borne Navy aircraft, but the Marine Corps played its part. Stationed on Midway Island was Marine Air Group 22 (MAG-22), comprising Marine Scout Bomber Squadron 241 (VMSB-241), equipped with obsolescent Vought SB2U Vindicator dive-bombers; and Marine Fighter Squadron 221 (VMF-221), flying equally vulnerable

Brewster F2A-3 Buffaloes and a handful of rugged, agile Grumman F4F-3 Wildcats. The Vindicators had flown direct to Midway from Hickam Field, Oahu, covering 1,137 miles in 9hrs and 20 minutes, at the time the longest massed flight of single-engined land 'planes. The Brewster Buffaloes had arrived at Christmas, flown from the deck of *Saratoga* after her abortive mission to relieve Wake Island. By the end of March, MAG-22 had received further reinforcements in the form of 16 Douglas SBD-2 Dauntless dive-bombers. On the eve of the Japanese attack on Midway there were 107 Marine, Army and Navy aircraft on Midway.

At 0555 hours on 4 June, Midway's radar logged the approach of "many planes" — a Japanese strike force consisting of 36 "Val" dive-bombers, 36 "Kate" torpedo-bombers and 36 escorting Zeros. Within 10 minutes MAG-22 — which already had fighters up to cover a sortie by the island's Consolidated PBY flying boats — had every available aircraft in the air. VMF-221, dividing into two groups, took off to intercept the incoming strike force. VMSB-241 headed for a rendezvous station 20 minutes to the east, where the dive-bombers' pilots were to receive further instructions.

The VMF-221 fighters, under Major Floyd B. Parks, sighted the Zero-escorted "Val" bombers at about 0616hrs, 30 miles from Midway. Captain John F. Carey, leading one of Parks' divisions in a Wildcat, flew into the attack from 17,000ft. His fighters had time for only one pass at the bombers before they were overwhelmed by the Zeros. Outclassed and outmaneuvered, nine of the 12 Marine fighters were shot down in the ensuing *mêlée*.

Like the fighter squadron, VMSB-241 had divided into two striking units, the first composed of 16 SBD-2s led by Major Lofton R. Henderson and the second of 11 SB2U-3s commanded by Major Benjamin W. Norris. Climbing to 9,000ft, Henderson's unit sighted the enemy carriers at 0744hrs, but were almost immediately set upon by Zeros flying cover. Only eight SBD-2s returned to Midway, having scored no hits. Henderson was among those lost.

Major Norris attacked the carriers in the wake of the bombing run made by 15 USAAF B-17s led by Lieutenant-Colonel Walter S. Sweeny. Roughly handled by the ever-present Zeros, Norris selected the nearest targets to hand, the battleships *Haruna* and *Kirishima*. No hits were scored and three of the SB2Us were shot down. One pilot who made it back counted 259 hits on his aircraft.

By 1100hrs all the surviving Marine aircraft had returned to Midway. Of the 25 fighters from VMF-221 involved in the action, there were only 10 left, and only two of these were fit to fly; 13 Buffaloes and two Wildcats were missing. Midway was covered in a pall of smoke from oil fires set off by the Japanese bombing, and ruptured fuel lines left two-thirds of the aviation fuel supply temporarily unavailable.

At 1700hrs a burning enemy carrier was reported 200 miles north-west of Midway. Major Norris prepared VMSB-241's six operational SBD-2s for a night attack. They took off at 1900hrs, but failed to locate the carrier. Norris did not return from the mission.

Left: An SB2U-3 Vindicator dive-bomber of VMSB-241 takes off from the airstrip at Midway.

CACTUS AIR FORCE

The original units of 1st Marine Aircraft Wing (MAW-1) to fly into Guadalcanal and lay the foundations of Cactus Air Force arrived on 20 August 1942 with Marine Air Group 23 (MAG-23). They were VMF-223, an F4F Wildcat squadron, and VMSB-232, an SBD Dauntless Squadron. On 22 August they were joined by the P-400 (a version of the Bell P-39 Cobra) aircraft of 67th Pursuit Squadron. The first of the Navy carrier detachments arrived from *Enterprise* on 24 August, and on the 30th the rest of MAG-23 landed in the form of VMF-224 and VMSB-231. On 3 September the command echelon of MAW-1 arrived on Guadalcanal, led by Brigadier-General Roy S. Geiger. During the period September–November 1942, many units served with Cactus Air Force for periods varying from three days to over six weeks. These detachments from Navy carrier groups included VS-3, VF-5, VT-8, VMSB-141, MAG-14, VS-71, VMF-121, VB-6, VMF-212, VMSB-132, VMF-112, VMSB-131, VMSB-142, VF-10, VB-10, VS-10 and VT-10. During the period 7 August–15 November, Guadalcanal-based aircraft (mainly SBDs) sank or wrecked 20 ships and damaged 14 more. Marine Wildcat pilots had claimed 407 victories, to which must be added 59 by Guadalcanal-based Navy squadrons and 18 by the Army's P-39s and P-400s, plus over 20 claimed by the bomber crews. Estimated Japanese losses to these units give a probable total of 260 aircraft. Cactus lost 101 aircraft, but of 126 Marine and 34 Navy fighter pilots who served on the island only 38 were killed or seriously wounded and, of these, 17 fell during the first week of combat. Outstanding aces of the 'Canal included Captain Joe Foss of VMF-121, the first American of the war to reach 20 victories, who gained the majority of his 26 kills at Guadalcanal; Major John L. Smith of VMF-223, who accounted for 19 enemy aircraft during the battle; and Lieutenant-Colonel "Indian Joe" Bauer of VMF-212 (11 victories), one of the great all-round fighter pilots of the Pacific war.

Guadalcanal

The Marine airmen at Midway had not died in vain. The gallant Henderson's name was soon to be immortalized when it was given to the jungle airstrip — Henderson Field — which became the focus of the bitter six-month struggle for Guadalcanal.

Guadalcanal is a jungle-carpeted island some 90 miles long and 25 miles wide in the Solomons group, which lies on the north-eastern approaches to Australia. Early in 1942 it assumed a critical strategic importance. If seized by the Japanese, Guadalcanal could be used to cut the sea supply routes between the United States and Australia. In American hands, it could serve first as a shield to the Allied build-up in Australia and then as a springboard for pushing the Japanese out of the South Pacific.

Before the Battle of Midway, the Imperial General Headquarters had authorized a series of ambitious operations in the islands

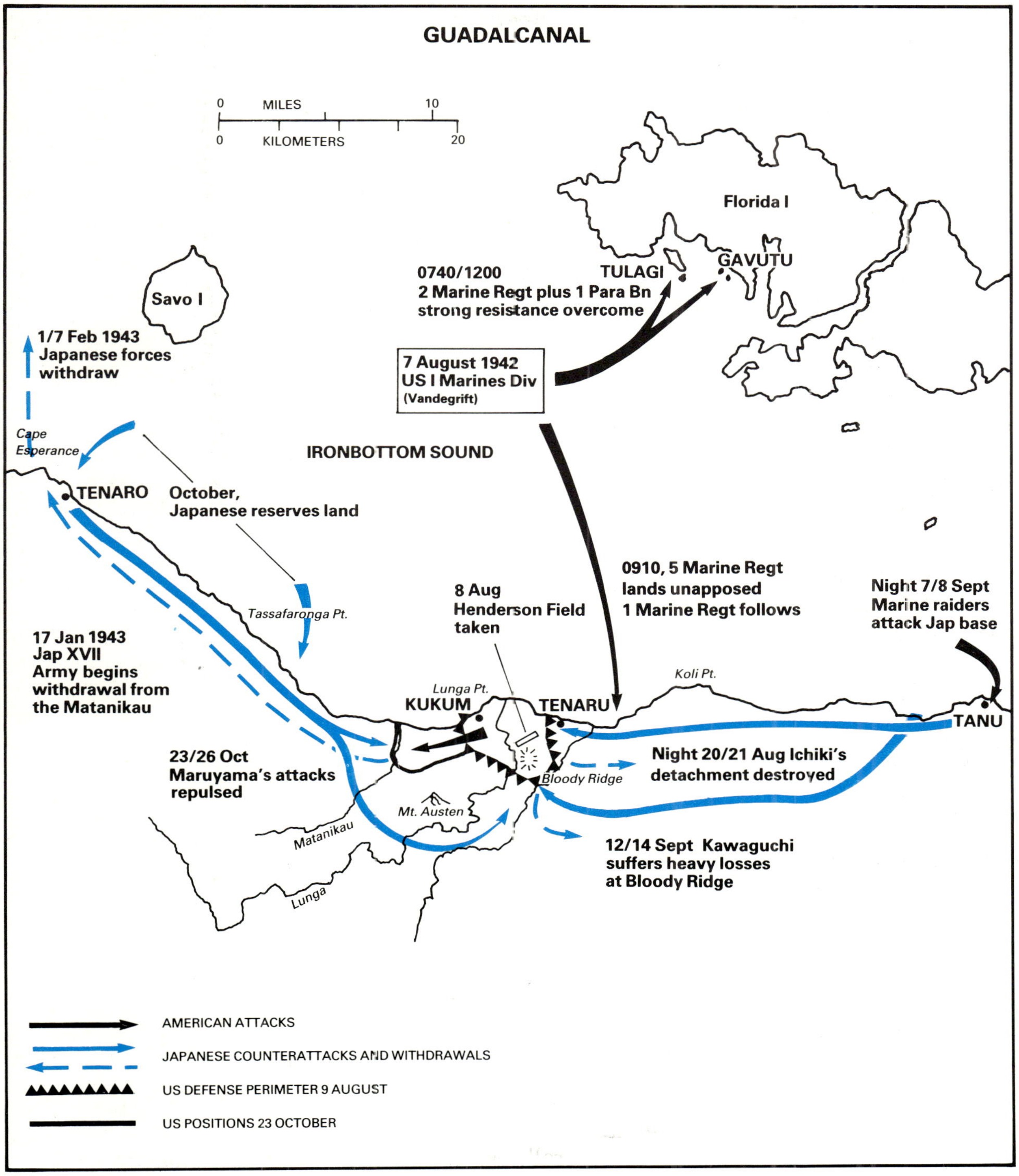

north and north-east of Australia, including the establishment of air and naval bases in the Solomons. After Midway, however, these plans were drastically revised and scaled down.

In May 1942, the Japanese had seized the island of Tulagi, with its magnificent anchorage, 30 miles north of Guadalcanal. At the end of the month they inserted troops on Guadalcanal. By mid-June, Australian coastwatchers were sending back reports that the Japanese were building an airstrip on the north coast of Guadalcanal between Tenaro and Kukum. When completed, it would enable Japanese aircraft to dominate the area. The number of Japanese troops on the island rapidly rose to about 2,400.

In the first six months of 1942 the United States' commitment to the "Europe First" policy had limited the men and materiel available for large-scale operations in the Solomons. Now Admiral Ernest J. King, Chief of Naval Operations and a powerful advocate of the allocation of greater resources

Opposite, top: A Marine F4F Wildcat takes off from the airstrip at Henderson Field, focus of the bitter seven-month struggle for Guadalcanal. With a maximum speed of around 330mph, the Wildcat was rather slow, but this was offset by its strength and maneuverability. The Wildcat was virtually the only Navy and Marine fighter facing the massive Japanese onslaught from Pearl Harbor until the arrival of the F6F Hellcat at the end of August 1943.
Opposite, bottom: Major-General Alexander A. Vandergrift, architect of victory at Guadalcanal, takes a back seat to Lieutenant-General Thomas Holcomb, the Corps Commandant, in December 1942.

to the Pacific, urged an operation to deny the Japanese the Solomons once and for all. In spite of their reservations, the Joint Chiefs of Staff gave their consent. Eight months after the humiliation of Pearl Harbor, the United States launched its first offensive against the Japanese in the Pacific — Operation Watchtower.

The US 1st Marine Division, reinforced by Marine Raider and parachute units, was tasked with landing in the Tulagi/ Guadalcanal area at the beginning of August and establishing a "permanent lodgement". The landing force reserve was to be provided by Marines of 2nd Marine Division. This bold plan belied the difficulties which lay ahead and the preparedness of the Marines for so important an operation.

Operation Watchtower was a stab in the dark. In the summer of 1942 there was not a single accurate map of Guadalcanal. There was no reliable information on the target area's tides and reefs, and no accurate assessment of the strength and dispositions of the enemy. These were, perhaps, the least of the problems facing 1st Marine Division's commander, Major-General Alexander A. Vandegrift.

Rapid expansion following Pearl Harbor had brought the division up to wartime strength, but thousands of its men were raw recruits with no experience beyond basic drill and weapons training. The division had not participated in any large-scale exercises and many of its units had no experience of amphibious maneuvers. When the division had been despatched from its base in North Carolina, General Vandegrift had been assured that his men would be given plenty of time for training and acclimatization as they were not likely to go into action until 1943. The first Vandegrift learned of the Marines' task was when he called on Admiral Ghormley of the US Navy, the commander of the South Pacific area and overall commander of the expedition, in New Zealand on 25 June. Half of Vandegrift's division had not yet arrived and D-Day — eventually fixed for 7 August — was just six weeks away.

The US Navy forces assembled for Watchtower were as inexperienced in amphibious operations as many of Vandegrift's Marines. A dress-rehearsal for the operation was held off Koro in the Fijis on 26 July. The aim was to test the fundamentals of an amphibious assault: air and naval gunfire support, air-ground communications, ship-to-shore communications, ground-controlled aerial bombing and disembarkation procedures. The result was a fiasco. Coral reefs prevented many assault craft from landing on their assigned beaches; many of the landing craft broke down; and the naval gunfire and close-air support was wildly inaccurate. Vandegrift could only hope that everything would be all right on the night.

In the early hours of 7 August, Martin Clemens, an Australian coastwatcher on Guadalcanal , was woken in his mountainside hideout by the sound of heavy guns. As dawn came up he had a grandstand view of the the landings on the island. The grey transport ships of the South Pacific Amphibious Force, commanded by Rear-Admiral Richmond K. Turner, moved deliberately toward their allotted positions off Guadalcanal and the neighboring islands. Simultaneously, the British Rear-Admiral Victor Crutchley deployed three Australian cruisers (*Australia*, *Hobart*, *Canberra*), two US cruisers (*Chicago*, *San Juan*) and nine destroyers north-east and north-west of Savo Island on the most likely Japanese avenue of approach. Over the horizon, Vice-Admiral Frank Fletcher's three carriers (*Saratoga*, *Wasp*, *Enterprise*) maneuvered 100 miles south of Guadalcanal. Screened by six cruisers and 16 destroyers, they provided air cover for the assault force.

The assault force had achieved total tactical surprise. The Japanese on Guadalcanal and the neighboring islands of Florida, Tulagi, Gavutu and Tanambogo did not know the enemy were on them until the first bombs and shells fell on their positions. At 0641hrs, after a 30-minute bombardment, the order was given to land the Marines on Guadalcanal. Thousands of tense, grim-faced Marines clambered down the cargo nets hanging from the decks of the transport ships and into the assault craft below. The landing craft churned through the calm sea to their assembly areas, forming into "boat groups" and closing with Red Beach in regulated waves. At 0910 hrs the first wave (1st and 2nd Battalions, 5th Marines) hit the beach on a 1,600-yard front. The landing crafts' hulls juddered into Guadalcanal's white sand, and the bombardment shifted inland as the Marines waded ashore through the warm, greenish-blue water, their weapons held high.

Pushing into jungle, they hastily established a defensive perimeter 600yds

inland. The Marines found it heavy going. Two weeks in cramped troop transports had sapped their stamina. Heavy packs, insufficient water and salt tablets, and the steamy unfamiliar jungle terrain, reduced many of them to a state of exhaustion. Advanced units had moved about a mile inland when General Vandegrift ordered a halt to re-orientate and establish internal contact. When night fell, the jungle reverberated with the sound of nervous sentries' fire against imaginary targets. General Vandegrift noted the "uniform and lamentable" failure of all units properly to patrol their fronts and flanks.

The Japanese failure to contest the landings on Guadalcanal gave Vandegrift a crucial breathing space. On Red Beach, 500

Above: Marines coming ashore at Guadalcanal take advantage of a short rest before moving into position.

men of 1st Pioneer Battalion struggled to prevent the logistical build-up from turning into a shambles. The landing area rapidly became clogged with a jumble of randomly dumped materiel through which the later waves of Marines wandered in search of orders. Off the shore, scores of LVAs sailed to and fro, searching for a clear space to leave their supplies. If enemy aircraft from Rabaul had struck hard at the congested beaches, Operation Watchtower might have failed at the first obstacle.

On the islands neighboring Guadalcanal — Tulagi, Gavutu and Tanambogo — the Japanese offered fierce resistance. Vandegrift had carefully picked his disembarkation point on Guadalcanal where there was no enemy. On Guadalcanal's smaller neighbors there could be no avoiding the Japanese, and the task of securing them had been given to the most seasoned Marine units.

Colonel Merritt "Red Mike" Edson, commander of 1st Marine Raider Battalion, tasked with the capture of Tulagi, ordered his men to strip down to the minimum combat equipment, telling them, "Don't worry about food. The Japs eat it, too. All you have to do is get it."

It took three days to crush the Japanese resistance on Tulagi, Gavutu and Tanambogo. Their defenders, fighting from coral caves, natural redoubts and heavily fortified emplacements, had literally to be blasted into submission. In the final assault on Tanambogo, the destroyer *Buchanan* was called in to pour shells into the last Japanese strongpoints at point-blank range. Resistance collapsed and within minutes the Marines had secured the island. The Marines had lost 144 dead or missing and almost 200 wounded. About 700 Japanese had been killed — about 90 percent of their number — many of them in mass suicide charges. This savage passage of arms gave the Marines a grim foretaste of the fanatical Japanese will to resist to the last man from fortified emplacements, and the extreme difficulty of assaulting heavily defended islands.

The first Marine patrols reached Guadalcanal's airfield late in the afternoon of the 8th. Everywhere were signs of the panic which had gripped the Japanese construction teams and the Naval Landing Force garrison when they had fled into the jungle after the initial American bombardment. Behind them they had left a large number of heavy vehicles, repair facilities, fuel dumps and quantities of food.

Satisfaction at the capture of the airfield was shortlived. From the outset, Admiral Fletcher had promised the land forces only two days' support, and he was now about to withdraw his carriers. At the same time a force of seven Japanese cruisers under Vice-Admiral Mikawa moved in to attack US transports unloading off Guadalcanal. On the night of 8/9 August, Mikawa encountered Crutchley's cruisers, which were patrolling the approaches to Savo Island. The Japanese had superior torpedo and night-fighting techniques, and in the engagement four of Crutchley's cruisers were sunk. The remaining elements of the naval force were now withdrawn, the last of Admiral Turner's ships sailing away on the afternoon of the 9th. The 11,145 Marines on Guadalcanal were on their own, with 50 percent of their equipment and supplies disappearing over the horizon in Turner's transports.

General Vandegrift was now in a perilous position, short of supplies and heavy equipment and holding a small enclave surrounded by dense jungle in which lurked the survivors of the original Japanese occupation force. The Marines had five units of fire and food for 37 days. (The original loadings of 60 days' supplies and 10 units of fire were, respectively, 33 and 50 percent below the 90-day and 20-unit levels considered normal for this kind of operation).

The Japanese enjoyed complete air superiority, and the Marines were subjected to constant air raids as they labored to complete the airfield, now known as Henderson Field. The waters around Guadalcanal were controlled by Japanese warships, which regularly shelled the Lunga Point beach-head. At one point, a Japanese cruiser landed a 200-strong advance echelon and supplies along the coast in broad daylight.

Using the heavy equipment abandoned by the Japanese, work was quickly completed on the airfield and a start made on two subsidiary strips. On 20 August, two squadrons of 1st Marine Aircraft Wing (MAW-1) flew into Henderson Field: VMF-223, an F4F Wildcat squadron under Major John L. Smith; and VMSB-232, an SBD Dauntless squadron commanded by Lieutenant-Colonel Dick Mangrum. These were the first elements in what soon became known as Cactus Air Force, "Cactus" being

Below: A detachment of Marines pause in a jungle clearing, December 1942. The Marine in the right foreground has added netting to his helmet for use in camouflaging himself in dense undergrowth. Conditions on Guadalcanal were punishing, with an exceptionally high number of men suffering from dysentery, malnutrition, virulent fungus infection, exposure and battle fatigue.

Above: A Marine Raider unit comes ashore at Guadalcanal in November 1942.

the codename for Guadalcanal.

Within hours of Cactus Force's arrival, the Japanese attacked from the east. Lieutenant-General Hyakutake, commanding XVII Army in the South Pacific, had been ordered to "eliminate" the American force on Guadalcanal, a task he assigned to Major-General Kawaguchi's XXXV Infantry Brigade. However, the brigade was not fully assembled, and the decision was taken to send in two echelons from the only units available, the regiment commanded by Colonel Ichiki and a Special Naval Landing Force.

On the night of 18/19 August, six destroyers under the command of Rear-Admiral Tanaka landed the 815 men of Ichiki's first echelon 20 miles east of Lunga Point. Japanese intelligence was extremely poor — Ichiki had been told that there were no more than 2,000 Marines on Guadalcanal — and their plans fatally compromised by a reckless contempt for the enemy.

At 0700 hrs on 19 August, one of Ichiki's units encountered a patrol from A Company, 1st Marines. A fierce fire fight left 31 Japanese and three Marines dead. Examination of Japanese bodies revealed detailed maps of the Marines' defences.

Disturbed by the Marines' discovery of his landings, Ichiki immediately moved on to the offensive. On the night of 20/21 August, he ordered an attack on the left flank of Vandegrift's defensive perimeter. The initial attack went in at 0310hrs as 200 men waded across the River Ilu with fixed bayonets and hurled themselves on the Marines' lines. Most of them were mowed down by small-arms fire and canister-firing 37-mm anti-tank guns of the 1st Special Weapons Battalion. A second bayonet charge met a similar fate, following which Ichiki withdrew into the jungle.

Although it had been grievously mauled, Ichiki's echelon regrouped. When daylight came, the Marines were subjected to harassing fire from across the Ilu. Vandegrift, anticipating the arrival of Japanese reinforcements, moved swiftly to deliver the *coup de grâce* to the survivors of Ichiki's formation. Led by Lieutenant-Colonel Leonard B. Cresswell, the 1st Battalion, 1st Marines, crossed the Ilu further upstream and swung north to envelop the Japanese.

By early afternoon, Ichiki's men had been encircled and the final bloody phase of "The Battle of the Tenaru" began to unfold. Bombed and strafed by Cactus Air Force and

pounded at short-range by artillery, the Japanese were squeezed back towards the sea. The last pockets of resistance were overcome by Vandegrift's light tanks, which remorselessly crushed the dead, dying and living beneath their tracks. In his report Vandegrift wrote that "the rear of the tanks looked like meatgrinders". The Marines had lost 34 men dead and 75 wounded. Ichiki had lost all but a handful of his élite troops. Escaping with a few men, he marched eastwards to Taivu, where he tore up and burned the regimental colors and then committed *hara-kiri*.

The Battle of the Tenaru was only the opening exchange in the see-saw battle for Guadalcanal. The arrival of Marine aircraft on Henderson Field had conferred an immediate tactical advantage on the Americans. On 25 August, Rear-Admiral Tanaka's transport group, carrying Ichiki's second echelon and screened by a light cruiser and destroyers, was pounced upon by Cactus Air Force's dive-bombers. Tanaka's flagship, *Jintsu*, and the transport *Kenryu Maru*, were both hit and set on fire. Further attacks by USAAF B-17s from Espiritu Santo sent one of his destroyers and the transport to the bottom. Tanaka, frustrated in his aim of reinforcing Guadalcanal, retired to the advanced base in the Shortland Islands, north-west of Guadalcanal.

Japanese movement in the waters around Guadalcanal was now restricted to the hours between dusk and dawn. Accordingly, Tanaka reverted to his original tactic of reinforcing Guadalcanal by using fast destroyers on nocturnal "Rat Runs". By the end of August the resourceful Tanaka's destroyers were operating with a clockwork precision which earned the American nickname of the "Tokyo Express".

To investigate the Japanese build-up outside his enclave, Vandegrift despatched Edson's Marine Raiders by sea on 7/8 September to make a reconnaissance in force of the enemy base at Taivu. The Raiders returned with vital intelligence of Japanese strength and intentions. The news that General Kawaguchi had moved off into the jungle with the bulk of his force was a clear indication that a second Japanese attack was imminent. Vandegrift anticipated correctly that the main Japanese thrust would be launched against the low 1,000yd ridge extending south of Henderson Field and parallel to the Lunga River, an inviting avenue of approach to the Marine airstrips.

He placed 1st Raider Battalion in this sector supported by the 105-mm howitzers of 5th Battalion, 11th Marines and elements of the Special Weapons Battalion. His reserve, 2nd Battalion, 5th Marines, was placed immediately to their rear.

Kawaguchi had planned a three-jawed envelopment of Vandegrift's enclave from the west, south and south-east, co-ordinated with air and naval support. To the normal difficulties associated with such a plan, he added the complication of cutting a trail over the steep jungle-clad ridges and gorges from the Tasimboko area to a point south of Henderson Field. The aim was to avoid observation, but what little Kawaguchi gained in surprise he lost in co-ordination and the stamina of the troops under his command. From the outset he lost control of the battle. His troops were exhausted by the march through the jungle, and communications broke down. Kawaguchi pressed on.

The Japanese showed their hand at mid-day on 12 September when bombers launched a heavy attack on the ridge south of Henderson Field. As darkness fell, a Japanese cruiser and three destroyers began to shell the ridge.

The shelling gave way to a series of probing attacks, but it was not until 2100 hrs on 13 September that the "Battle of Bloody Ridge" reached a crescendo. Without artillery preparation, Kawaguchi drove a two-battalion charge against the center and right of the Raider-Parachute line. Marine mortars, sited in defilade, poured shells into the human waves as fast as their loaders could slide them down the tubes. Behind them the artillery on the ridge laid down a devastating fire, some of it falling within 200yds of the Marines' lines. Japanese mortar fire began to drum the ridge and at 2230hrs, after a violent barrage, they shifted their attack to the thin flank held by the Parachute troops. Screaming in English, "Gas Attack! Gas Attack!", they swarmed from the jungle through a smoke screen, driving the Parachute troops back along the ridge. The Marines' defenses buckled but did not break. The Raiders' communications with the artillery in their rear and Vandegrift's headquarters were severed. But Japanese flares "telegraphed" each successive attack, providing 11th Marines' gunners with reference points for the all-night firing in which they expended 1,992 rounds of 105-mm projectiles, some at ranges as short as

1,600 yards.

At one point the Japanese surged to within 1,000 yards of Henderson Field in some of the most savage fighting of the Guadalcanal campaign. They were thrown back, only to launch two more attacks before daybreak. But their strength was now ebbing away. When cannon-firing fighters from Henderson Field began strafing the jungle at the foot of the ridge, Kawaguchi gave the order to withdraw. He had lost over 1,200 men killed, wounded and missing. Demoralized, disease-ridden and starving, the survivors of XXXV Brigade straggled away westwards to link up with other Japanese units at Port Cruz.

The Marines had also taken heavy losses. Of just over 750 Raiders who had landed on 7 August, 234 had become casualties; of the

The debris of battle. A Japanese landing barge which never reached its destination, wrecked by fire from Marine half-tracks during the Guadalcanal campaign.

377 Parachute troops, 212 had been killed

The response of Imperial General Headquarters was to despatch the veteran XXXVIII Division to the South Pacific to join XVII Army. Lieutenant-General Maruyama's II Division — already in the area — was ordered to Guadalcanal. Japanese operations on New Guinea were suspended in order to facilitate a concentration of all available naval, air and military resources on the recapturing of Guadalcanal. General Hyakutake transferred his headquarters to the island, where he had direct control of 20,000 men, including a regiment of three batteries of heavy artillery, a mortar battalion and a tank company.

The Marines also received much-needed reinforcements: 7th Marines, an artillery battalion, motor transport companies, communications personnel and 164th Infantry Regiment. This brought the number of US troops on Guadalcanal to 23,000, but many of them were suffering from malnutrition, dysentery, virulent fungus infection, exposure and battle fatigue. In other less crucial battle zones, more than one-third of them would have been invalided out.

Cactus Air Force was also stretched to the limit. Japanese attempts to feed in reinforcements led to a naval battle off Cape Esperance at the northern end of Guadalcanal. On the night of 11/12 October, a Japanese convoy sailing down the "Slot" — the narrow channel inside the Solomons chain — clashed with the cruisers commanded by Rear-Admiral Norman Scott. After a confused and ferocious engagement, the Japanese withdrew, although the transports they were escorting got through to Guadalcanal, landing heavy artillery and other reinforcements unopposed.

At first light on the 12th, all available SBDs, F4Fs and P-39s took off in search of any cripples from the night engagement. They were led by Colonel Al Cooley, whose MAG-14 had relieved MAG-23, and Lieutenant-Commander John Eldridge, commanding VS-71. The attack group found the survivors north of the Russell Islands, sinking two destroyers and inflicting heavy damage on the remaining enemy ships.

During the night of 13/14 October, the battle ships *Kongo* and *Haruna* sailed close into the Guadalcanal coastline and delivered a 90-minute bombardment of the Marines' positions with their 14-in guns. Over 900 shells were fired into the beach-head, most of them landing on Henderson Field and in the coconut grove at the north-eastern side of the strip, where the pilots and crews were billeted. Forty-one officers and men were killed; of the 38 SBDs operational 24 hours before, only seven remained fit for flying. The Field was pockmarked with huge craters and strewn with blazing wreckage but, miraculously, the fighter strip was virtually unscathed and 29 of Cactus Air Forces' 42 F4Fs were still in flyable shape.

Air raids on the 13th and 14th inflicted further damage, and another bombardment from the battleships in the small hours of the 15th left so few aircraft serviceable, and petrol in such short supply, that a Japanese convoy of six transports — sailing in broad

Above: Lieutenant Mitchell Paige of 7th Marines receives the Medal of Honor from General Vandegrift for gallantry during the fighting on Guadalcanal. The US Marines who served on the 'Canal composed a little verse which vividly conveys their feelings about the campaign: "And when he gets to Heaven, To St Peter he will tell: 'One more Marine reporting, Sir — I've served my time in Hell!' "

daylight down the Slot — was able to put 3,500 troops ashore at Tassafaronga.

In a supreme effort, Cactus Air Force put 12 SBDs and eight fighters in the air for an attack on the enemy transports. Flying with them was a Consolidated PBY Catalina flying boat piloted by General Geiger's aide, Major Jack Cram, and rigged with two torpedoes. In one of the most remarkable torpedo attacks in military history, Cram glided in from 6,000ft, releasing the torpedoes at 200ft and scoring a direct hit on one of the transports. The lumbering Catalina flew back to Henderson Field with Japanese fighters swarming all over it, but Cram made it back with the help of a landing F4F, whose pilot shot down the last of his pursuers without bothering to pull up his gear.

On 23 October the Japanese II Division — some 5,600 men — attacked the western perimeter of Henderson Field across the Matanikau River. The Marines were well dug in and ready for them, smashing the attack with concentrated artillery fire. Twenty-four hours later, General Maruyama launched the main thrust from the south with more than 7,000 men. For two days the Japanese force battered itself to destruction on the ridges to the south of Henderson Field. Then, like Kawaguchi's brigade, they fell back into the jungle, having lost 3,500 men.

Poor co-ordination had doomed Maruyama's offensive. His arduous approach march through the jungle had held him up and forced him to abandon much of his artillery. Twice he postponed his attack, but news of the second postponement failed to reach the commander on the Matanikau. The plans for a simultaneous two-pronged assault on Henderson Field were destroyed, and with them the dwindling hopes of victory on Guadalcanal. Operating on interior lines, Vandegrift had defeated each attack in turn. American losses had been fewer than 350, and they now began to expand the perimeter.

Maruyama's attack on Henderson Field was to have been supported by aircraft of the Combined Fleet, stationed north of Guadalcanal. Simultaneously, US Task Forces 16 (with the carrier *Enterprise*) and 17 (with the carrier *Hornet*) were ordered to patrol the area around the Santa Cruz islands to intercept Japanese forces approaching Guadalcanal. The fleets clashed on 26 October. The Japanese carrier *Zuiho* was disabled by dive-bombers from *Enterprise*, which in turn was hit by air strikes. *Hornet* was swamped by dive- and torpedo-bomber attacks and reduced to a blazing hulk which was later sunk by the Japanese. The battle was broken off with the disabled enemy carrier limping away in company with *Zuiho*. Having lost *Hornet*, and with *Enterprise* out of action, there was not a single carrier operational with the Pacific Fleet. But the Japanese had won a Pyrrhic victory, having lost about 100 aircraft and their irreplaceable experienced aircrews.

Refusing to accept defeat, the Japanese continued to pour men and supplies into Guadalcanal. But in mid-November 11 transports carrying the bulk of Lieutenant-General Sano's XXVIII Division to Guadalcanal were intercepted by aircraft flying from Henderson Field. Seven of the transports, and 4,000 men, were lost. Only 2,000 men, most of them without equipment, reached Guadalcanal. Thereafter General Hyakutake was isolated; the supplies and reinforcements reaching him shrank to a trickle.

Preparing to go on the offensive, the Americans brought in fresh units to replace the war-weary 1st Marine Division, which was relieved at the beginning of December by the 25th Infantry Division, 2nd Marine Division and the Americal Division, designated 14th Corps under the command of General Alexander Patch. At the end of the month, Emperor Hirohito authorized the withdrawal of XVII Army from Guadalcanal. In a text-book evacuation during the first nine days of February, the destroyers of the "Tokyo Express" took off the 12,000 exhausted survivors.

Overall Japanese casualties in the fighting on Guadalcanal were 14,800 killed or missing in action, 9,000 dead from disease and 1,000 taken prisoner. Marine and Army casualties

MARINE RAIDERS AND PARACHUTE UNITS

In the 1930s, the Marine Corps experimented with the idea of using raider teams to operate behind enemy lines. The concept achieved formal recognition in January 1942 when 1st Battalion, 5th Marines, was redesignated 1st Separate Battalion. A month later a Marine officer, Lieutenant-Colonel Evans F. Carlson, was ordered to raise men to form the 2nd Separate Battalion. Lieutenant-Colonel Merritt A. Edson was appointed commander of 1st Battalion. Within weeks the units were redesignated 1st and 2nd Raider Battalions and readied for combat. Each battalion was made up of 850 officers and men, divided into six rifle companies and a small headquarters. In addition to standard infantry equipment, Raiders used shotguns, bangalore torpedoes, chainsaws and rubber dinghies. They had three specific roles: spearheading amphibious landings; launching hit-and-run raids against enemy-held islands; and conducting protracted guerrilla operations behind enemy lines. By mid-1942 both Raider Battalions had been blooded. In August, Carlson's Raiders hit Makin Island in a lightning raid, and Edson's men were tasked with the taking of Tulagi in the initial phase of the battle for Guadacanal. The success of the Raiders led to the formation of more units, which were formed into 1st and 2nd Raider Regiments.

The first Marine parachute unit was raised in 1941, taking part in combined US Army Corps/Marine Corps exercises at Camp Lejeune that year. Subsequently two more companies were raised, and as 1st Parachute Battalion, Fleet Marine Force, the unit fought on Tulagi, Gavutu and Guadalcanal. Three more parachute battalions were then formed. Early in 1943, 2nd and 3rd Battalions joined 1st Battalion as 1st Parachute Regiment; 4th Battalion was never deployed as a unit from the United States. In spite of their name and training, none of the parachute units ever jumped in combat. The Parachute Regiment (minus 2nd Battalion) fought at Bougainville in November 1943. The 2nd Battalion participated in the operation in Vella Lavella under the command of 8th New Zealand Brigade, suffering heavy casualties during the initial landings when the ship carrying the battalion HQ was sunk. In October 1943 2nd Battalion was in action in the diversionary operation against Choiseul. The Parachute Regiment was disbanded in 1944 and its units merged with 5th Marine Division, fighting at Iwo Jima in 1945.

totalled 1,598 officers and men killed and 4,709 wounded. Of these, Marine ground forces dead were 1,152 and 2,799 wounded.

For the Allies, the seizure and successful defense of Henderson Field brought crucial strategic gains. Australia and New Zealand were now safe, and Allied forces stood on the flank of the Palau–Truk–Marshalls line, the outer cordon of the Japanese empire.

The battle for Guadalcanal was also a psychological victory over the previously all-conquering Japanese Army. The Marines had demonstrated that the Japanese could be beaten on their own ground. Not the least of their achievements was the 150-mile trek behind enemy lines undertaken by 2nd Marine Raider Battalion under Colonel Evans F. Carlson. Landing at Aola Bay on 4 November, Carlson's Raiders fought their way through the jungle, in four weeks killing over 500 Japanese in a dozen actions for the loss of only 17 men and driving the enemy deep into the interior of Guadalcanal. No longer could the Japanese claim absolute supremacy in the jungle.

Above all, Guadalcanal was a primer of amphibious and jungle tactics. As the official history of the US Marine Corps observes, "It was everything the United States could do at the moment against everything the Japanese could manage at that place". General Vandegrift summed up the importance of the campaign in characteristically pithy fashion: "We struck at Guadalcanal to halt the advance of the Japanese. We did not know how strong he was, nor did we know his plans. We knew only that he was moving down the island chain and that he had to be stopped. We were all well-trained and as well-armed as time and our peacetime experience allowed us to be. We needed combat to tell us how effective our training, our doctrines, and our weapons had been. We tested them against the enemy, and we found that they worked. From that moment in 1942, the tide turned, and the Japanese never again advanced."

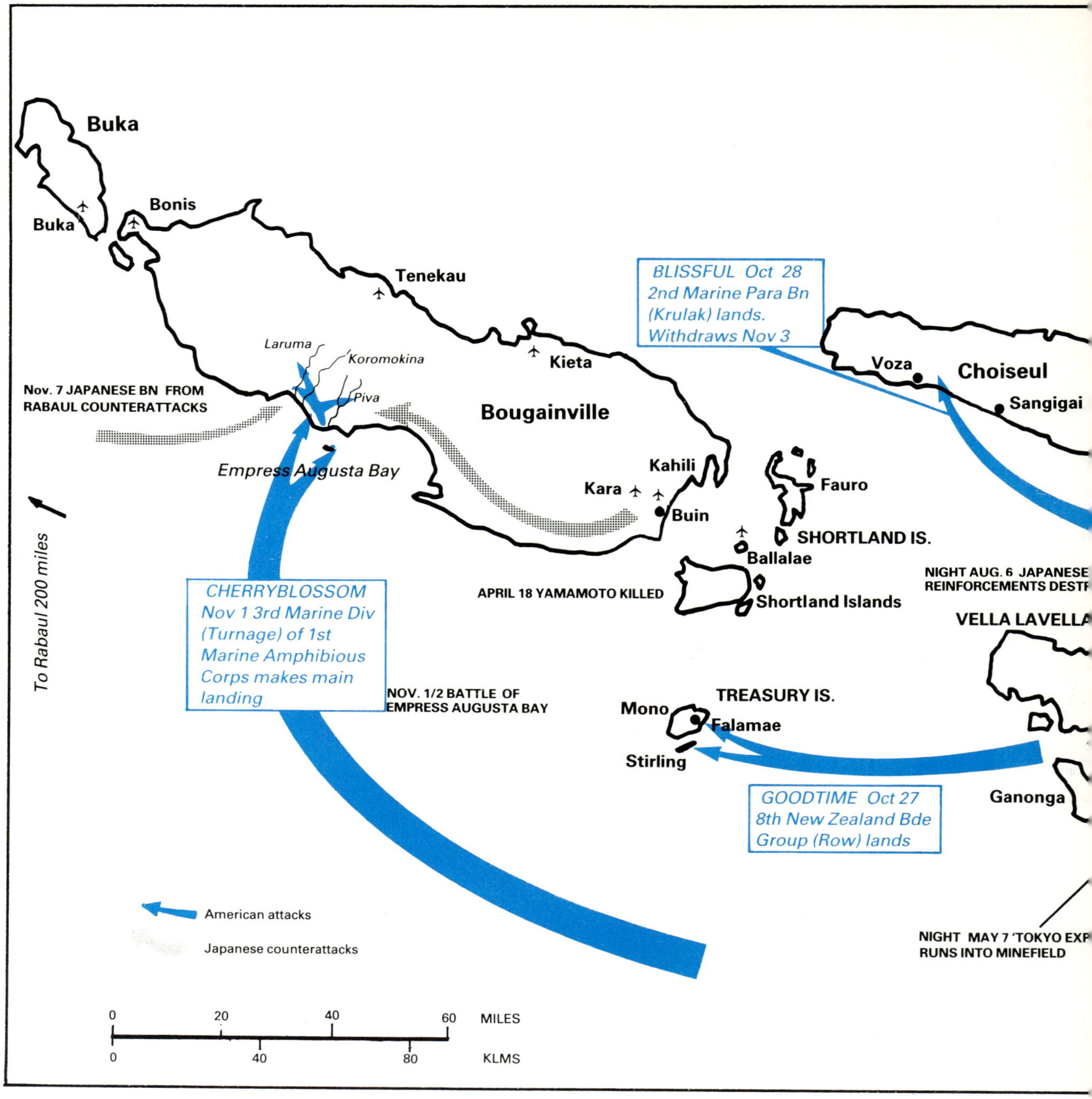

The Battle for the Solomons

Guadalcanal was a setback for the Japanese. But they had anticipated a shift to a defensive strategy once their perimeter had been anchored, and the defeats at Midway and Guadalcanal had taken place beyond the limits of the perimeter as they had originally envisaged it. They were more concerned with the alarming attrition of trained carrier pilots. They still nursed hopes that a series of defeats on the main line of their perimeter would force the Americans to give up.

On the Allied side, the victory at Guadalcanal marked the beginning of the Pacific drive on Japan. General MacArthur, Supreme Commander South-West Pacific, was to outflank the Japanese base at Rabaul and approach the Philippines from the south. Admiral Halsey, commanding the South Pacific area, was to co-operate with MacArthur in isolating the fortress of Rabaul and driving north-west from Guadalcanal along the Solomon Islands. Admiral Nimitz, commanding the Eastern and North Pacific areas, was to channel resources to the other two commanders while launching a central drive through the Pacific. The axes of the final assault on Japan were to run from the Philippines through Okinawa; from the Marianas through Iwo Jima; and southwards from the Aleutians.

In the South-West Pacific area, units of the US Army would shoulder the burden of the fighting. The Marines would carry the landings in the Central Pacific. To meet this challenge, the Corps was to be expanded to 300,000 officers and men by mid-1943. This rapid expansion led to the abandonment of the Corps' policy of segregation and the admission of blacks into its ranks, a reform

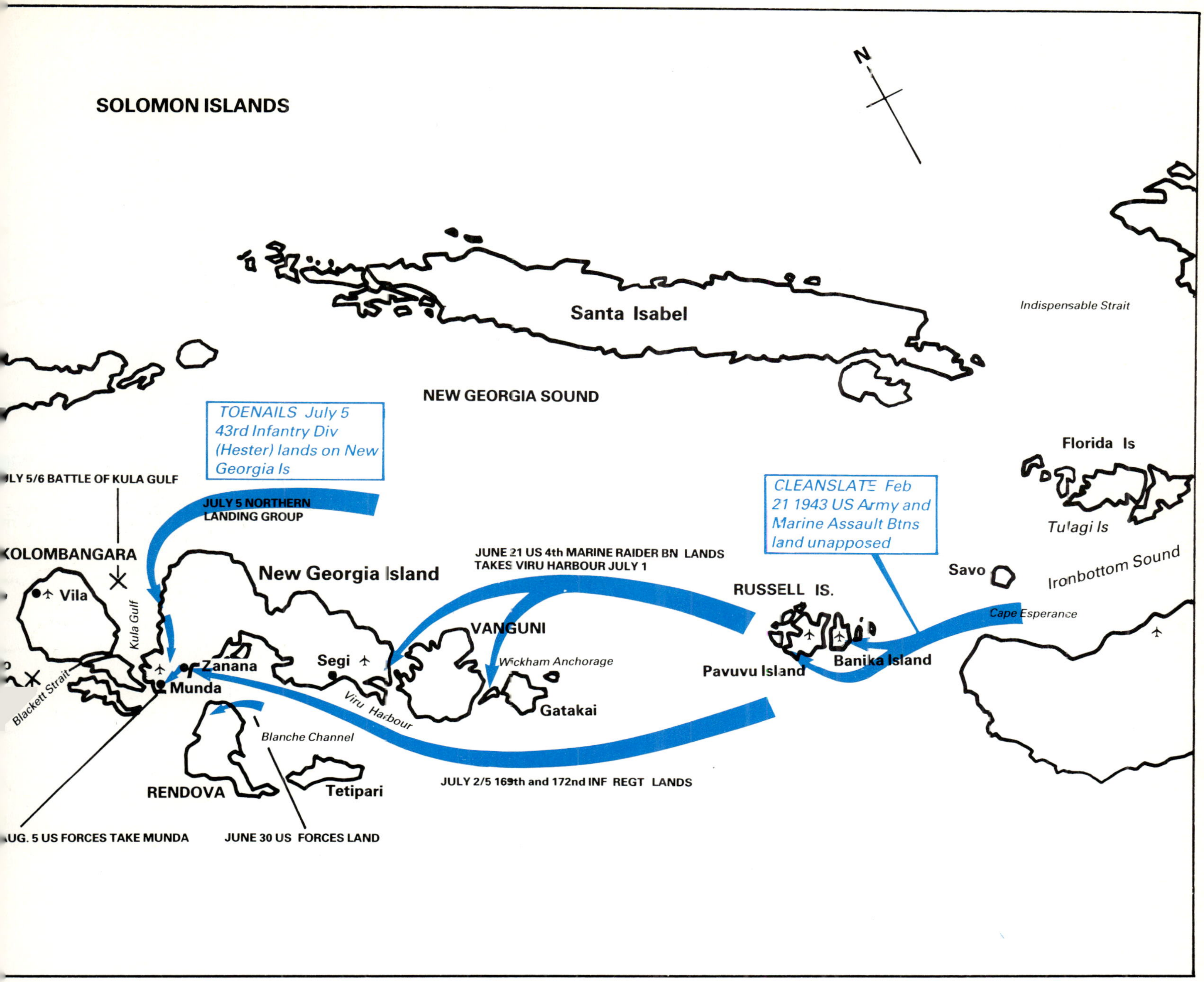

initially undertaken with the greatest reluctance. At first blacks were confined to segregated units, under white officers and NCOs, and given only defense and labor duties. Black Marines were eventually introduced to combat, and by the end of the war over 15,000 had passed through.

For the Allies, the Guadalcanal campaign had been only the beginning of the battle for the Solomons. Further advances along the island chain to seal off Rabaul from the south-east set the pattern for subsequent operations in the Pacific. Key air bases were seized by outflanking moves, to provide an overlapping system of controlled air space. Intermediate enemy-held islands were mopped up while preparations were made for the next advance. Principal objectives in the Solomons were the New Georgia group and Bougainville.

Between February and June 1943, there was a breathing space while Admiral Halsey prepared for the assault on New Georgia. On 21 February, in Operation Cleanslate, the Russell Islands — between Guadalcanal and New Georgia — were seized in an unopposed operation. The capture of the Russell Islands provided the Americans with sites for fighter bases to cover movements through the central Solomons. Further support came from new airstrips constructed on Guadalcanal.

On 18 April Admiral Yamamoto, Japan's foremost strategist, was intercepted and shot down by US fighters while flying from Rabaul to Bougainville. At the end of the month, General MacArthur issued his final plan for Operation Cartwheel, a co-ordinated drive in the South and South-West Pacific towards Rabaul.

The campaign in the central Solomons was

Following page: Marine Raiders and their dogs on Bougainville.
Below: Marine infantry and light tanks encounter heavy going in the jungle on Bougainville.

opened on 21 June when 4th Marine Raider Battalion landed at Segi on the eastern tip of New Georgia. Principal target on New Georgia was the airfield at Munda.

The Army units tasked with taking Munda — 169th and 172nd Infantry Regiments and 43rd Infantry Division — were reinforced with two Marine Raider and two Defense battalions. The Marines were also assigned a role in the capture of Rendova, the island to the south of Munda, from which artillery support could be provided for the assault on the airfield. Between 2 and 5 July, troops were landed to attack Munda from north and south. As this was the most important airfield in the Solomons, it was heavily defended and was not taken until 5 August. Marines played a vital role in beating off successive Japanese counter-attacks around the airfield and provided artillery support from Rendova with their new 155-mm "Long Toms". They also supplied six of the 32 fighter and bomber squadrons flying out of Guadalcanal and the Russells during Operation Toenails. On the night of 6 August, Japanese reinforcements were destroyed in a naval battle in the Vella Gulf. By 22 August the enemy had evacuated northern New Georgia, and three days later all resistance on the island had ceased.

In mid-August, a US Combat Team had

landed on Vella Lavella, west of New Georgia, and established an airbase which leapfrogged the intermediate island of Kolombangara. New Zealand troops replaced the US forces on Vella Lavella in mid-September and cleared the island of the remnants of the Japanese garrison. This forced the enemy to abandon Kolombangara, and by the end of the first week in October the campaign in the central Solomons had been brought to a successful conclusion.

On 1 October Admiral Halsey ordered an invasion of the island of Bougainville. Its capture would place the Japanese base at Rabaul within range of American fighters and dive-bombers. However, it presented a formidable obstacle. Approximately 125 miles long, Bougainville offered a nightmarish prospect for military operations, carpeted with swamps and jungle from which rose the jagged peaks of a mountainous spine. A line of Japanese airfields stretched from Bonis, on Bougainville's northern tip, to Buin in the south. Defending them were 40,000 troops.

To disguise their intentions, the Allies launched two diversionary operations. In Operation Blissful, on 28 October, 2nd Battalion, 1st Marine Parachute Regiment under Lieutenant-Colonel Victor H. Krulak, made a feint assault on the island of Choiseul, to the north of New Georgia. Simultaneously, in Operation Goodtime, the 8th New Zealand Brigade landed on the Treasury Islands, south-west of Vella Lavella.

These feints successfully distracted Japanese attention. When the attack on Bougainville — Operation Cherryblossom — opened on 1 November, 3rd Marine Division met only light resistance as it went ashore at Cape Torokina on the island's southern coast. By nightfall, 14,000 troops had been put ashore.

Above: Marines pinned down on the beach at Bougainville. *Top:* A flamethrower is turned on a Japanese pillbox on Russell Island.

Above: Bougainville, November 1943. Casualties disembark from a tank landing craft after being ferried across Empress Augusta Bay to Puruata Island, where another LST waits to evacuate them.

One of the principal lessons of Guadalcanal had been that it was cheaper and easier to build a new airfield than to capture and then improve one the Japanese had already built or were building. This also coincided with one of the fundamentals of amphibious warfare — never hit a defended beach when the objective can be reached over an undefended one.

At first the Japanese commander on Bougainville, the ill-fated General Hayakutake, believed that the 1 November landings were another diversion, holding back his forces while the Americans began to dig in and construct an airfield. A Japanese counter-attack, launched from Rabaul on 7 December, failed to dislodge the American assault force. By the end of the year they had established a defensive perimeter stretching 5,000 yards inland and had turned Empress Augusta Bay into an Allied naval base with three airstrips in operation. Even though much of Bougainville remained in enemy hands, Admiral Halsey was now poised to close the pincers on Rabaul.

By the time the Japanese launched a belated March offensive against the American perimeter, 3rd Marine Division had been replaced by the Americal Division. The Japanese threw themselves on the carefully prepared American positions, and in 17 days of bitter fighting suffered 7,000 casualties; American losses were 1,000. By the end of April, US aircraft from the Torokina strips were joining carrier-based 'planes in regular missions against Rabaul.

The trap was closing on Rabaul. MacArthur's next move was the capture of Cape Gloucester, on the western end of New Britain Island, which controlled the Vitiaz and Dampier Straits between New Britain and New Guinea. With the straits in American hands, the sea lanes to the north-west would be safe for passage. In a three-week operation beginning on 26 December, elements of 1st and 7th Marines overcame 10,000 Japanese troops in and around Cape Gloucester. The capture of the Admiralty Islands and the St Matthias Group northwest of New Britain (February–March 1944) completed the encirclement of Rabaul, whose 135,000-strong garrison was left to "wither on the vine".

Tarawa

During the summer and autumn of 1943, before the pincers finally closed on Rabaul, Admiral Nimitz assembled an armada of ships, men and aircraft for the atoll-by-atoll drive in the Central Pacific. The offensive was designed to secure MacArthur's right flank and to provide forward bases for a strategic bombing campaign and a jumping-off point for landings on Japan itself. The three stepping stones in Nimitz's Central Pacific drive were the Gilbert Islands, the Marshalls and the Marianas.

The island chains of the Central Pacific presented operational problems markedly different from those which had been encountered at Guadalcanal and on Bougainville. They were made up of small atolls, rings of islets which were the tops of submerged mountains. These specks in the ocean had formidable natural defenses in the razor-sharp reefs which uniformly surrounded them. Their low contours offered little or no cover for amphibious assault forces coming ashore. Secure in fortified emplacements and bunkers, their Japanese defenders could survive the heaviest air and naval bombardments.

On 20 November 1943, the Central Pacific campaign opened with the launching of Operation Galvanic, the taking of the Gilberts, on the outer edge of Japan's Pacific empire. The forces assembled for Galvanic comprised the US 5th Fleet, under Vice-Admiral Raymond Spruance, and V Amphibious Corps, under the overall command of Major-General Holland M. "Howlin' Mad" Smith, a Marine veteran of the Aleutians campaign and one of the founding fathers of modern amphibious warfare.

By 23 November, Makin Island — the

Below: On the beach at Tarawa. Heavily-laden Marines climb the sea wall to attack Japanese positions around the atoll's airstrip.

scene of a daring Marine Raiders operation in August 1942 — had been secured by an Army landing force consisting of 165th Regimental Combat Team (reinforced) of 27th Division. But the toughest objective in the campaign lay ahead — Tarawa, a bracelet of tiny coral islands which was of crucial strategic importance. On one of its islets — Betio — there was a heavily defended airstrip suitable for fighters and light bombers. No thrust north-west to the Marshalls and Marianas could be undertaken until Tarawa had been seized.

Betio was only a few feet above the water, containing a scant 300 acres of sand, coral and palm trees. The Japanese had crammed 4,800 troops of the Special Naval Landing Force on Betio and added to the natural barrier of its twin coral reefs with a fearsome network of man-made defenses. The waters offshore were laced with mines, concealed obstacles and barbed wire, aimed at channelling attackers into prepared killing grounds. At the edge of the water a four-foot-high sea wall was defended by machine-gun posts with interlocking fields of fire.

Coastal defences included four 8-in guns, several smaller pieces, and a network of interlinked bunkers reinforced with concrete roofs up to 6½ft thick. Running out to sea were the island's coral reefs. Unknown to the Americans, the inner reef rose sufficiently high to ground their landing craft. It was planned to send only the first three waves of Marines ashore in LVTs, which could negotiate the barrier of the reefs. The remainder would have to carry the assault in LCVPs (Landing Craft Vehicle, Personnel) and LCMs (Landing Craft, Mechanised), both of which were likely to go aground on the edge of the reef, several hundred yards offshore.

The bombardment of Tarawa had begun on 13 November with a series of air strikes. Neither these, nor the 16-in shells of the battleships *Carolina* and *Maryland*, succeeded in suppressing the Japanese defenses. H-Hour was planned for 0830 hrs

on 20 November. The assault force was 2nd Marine Division, under Major-General Julian Smith. The attack was to be spearheaded by Colonel David Shoup's 2nd Marines and the 2nd Battalion, 8th Marine Regiment, with the remaining two battalions of 8th Marines being held back as a divisional reserve. Additionally, a corps reserve of three battalion landing teams of 6th Marines was held to cover the entire operation in the Gilberts, to be released to the division on request. Three landing areas were designated on Betio: Red Beaches 1, 2 and 3 on the island's lagoon side. The blunt western end of the islet, where no landings were planned, was designated Green Beach.

As the first wave of LVTs crossed the reef, the fire support lifted. Gazing at the dense cloud of smoke and coral dust swirling above Betio, the naval gunfire support commander, Admiral Kingman, thought it "impossible for any human being to be alive on Betio Island". But the enemy were ready and waiting.

Only the destroyers in the lagoon were close enough to give accurate suppressive fire during the Marines' first run-in to Betio. Thus the island's defenders had a breathing space in which to reach their posts and open up with a storm of machine-gun fire and 75-mm airburst shells.

The first Marines ashore, at 0855 hrs, were men of the scout and sniper platoon of 2nd Marine Regiment, led by Lieutenant William Hawkins, who took the end of the long pier which ran out from the corner of Red Beach 2. Behind Hawkins, the LVTs were taking heavy punishment. Half of them were destroyed in the water. Many of the remainder landed on the wrong beaches. None of them could surmount the sea wall. The Marines were pinned down on a few square yards of black sand in front of the barricades. There was no way back across the blood-tinged waters of the lagoon, and seemingly no way forward. The landing craft in the following waves were grounded hundreds of yards from the shore by an unexpectedly low tide, forcing the Marines to

LVTs churn towards the shore at Tarawa, 20 November 1943.

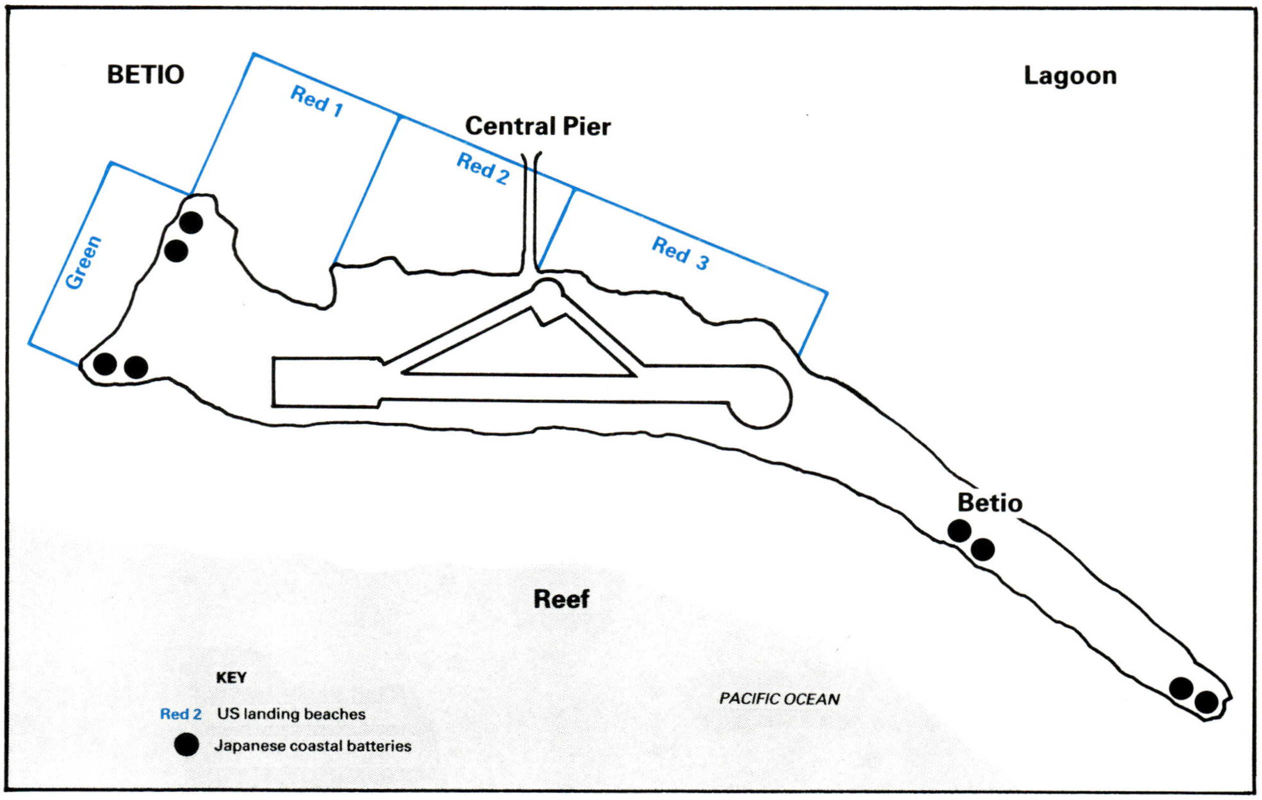

Fire-shredded palms, abandoned equipment and bloated corpses bear testimony to the savage fighting at the water's edge of Red Beach 2 during the landings on Tarawa. This was the objective of 2nd Battalion, 2nd Marines, under Colonel Herbert Amey. He and many of his men were cut down by withering fire from carefully hidden enemy bunkers.

wade in through the fire-lashed lagoon, past the bobbing corpses of their comrades.

The 3rd Battalion, 2nd Marines, tasked with landing on Red Beach 1, was virtually annihilated, its survivors forced to the western edge of Betio by furious Japanese fire from strongpoints dominating the landing area. On Red Beach 2, 2nd Battalion, 2nd Marines, took heavy casualties coming ashore, and at noon was reinforced by 1st Battalion, 100 of whose men were forced off course by enemy fire, joining the remnants of 3rd Battalion clinging to the western tip of Green Beach.

Only on Red Beach 3 was a firm lodgement made, by Major Henry Crowe's 2nd Battalion, 8th Marines. Accurate covering fire from the destroyer *Ringgold* ensured that casualties did not rise to 10 percent until the arrival of the fourth and fifth waves. By mid-day Crowe had a platoon of medium tanks available. Covered by 37-mm guns, his men began to push inland with the tanks and demolition teams in close mutual support. Soon afterwards Crowe's forces made contact with 1st and 2nd Battalions, 2nd Marines, on Red Beach 2, where Colonel David Shoup had placed his command post next to an enemy strongpoint. Shoup later recalled: "I was never off my feet for 50 hours, standing for most of the time by an enemy pillbox with 26 live Japs therein."

Having linked up, Shoup and Crowe's forces mounted an attack, supported by medium tanks, across the airstrip taxiway, establishing a joint perimeter some 300 yards deep.

The approach of darkness brought with it fears of a Japanese counter-attack, and 3rd Battalion, 8th Marines, was now committed to Red Beach 2, taking heavy casualties as they came in on both sides of the pier. On the extreme right flank of Red Beach 1, some scattered elements of 1st, 2nd and 3rd Battalions, 2nd Marines — driven to the western edge of Betio by fire from a chain of enemy strongpoints in the cove that comprised most of Red Beach 1 — launched an attack southwards along Green Beach at 1630 hrs. Supported by two tanks, they advanced about 500 yards before being brought to a halt. The tanks were knocked out and they lacked the flamethrowers and heavy equipment to deal with the network of Japanese bunkers which held them up. They withdrew to a defensive position for the night at the north end of Green Beach.

There was no Japanese counter-attack that night, probably because their communications had been wrecked by the naval bombardment and their strongpoints were unable to co-ordinate their actions or assess the overall situation.

Before dawn on the 21st, General Smith committed the remaining divisional reserve, 1st Battalion, 8th Marines, who had spent an uncomfortable night offshore in their LCVPs. On the long walk ashore they were badly mauled, losing 350 of their 800 men before joining the right flank of Shoup's forces in the centre of the island. An artillery battalion of 10th Marines had also arrived with five 75-mm pack howitzers, but could make no impression on the enemy defenses of Red Beach 1. There seemed little hope of joining hands with the Marines on Green Beach.

But during the morning two medium tanks and the vital flamethrowers arrived on Green Beach. Air strikes were called up to pulverize the Japanese batteries on Betio's western coast. At 1100 hrs the Marines advanced southwards, supported by naval gunfire. Green Beach was secured an hour later,

Right: First Lt. Marvin Rankin and Cpl. Southard plot the situation on Roi Island while under enemy fire, February 1944.

opening the way for the landing of reinforcements in battalion strength with minimum casualties and with their organization, command and equipment intact. The corps reserve, 6th Marines, was now on its way, and by late afternoon its 1st Battalion was coming ashore. The 2nd Battalion, 6th Marines, was deployed with artillery from 10th Marines on the neighboring island of Bairiki to add its firepower to the final assault and cut off a Japanese retreat across the narrow stretch of water separating the two islands.

The tide in Tarawa's lagoon shifted, allowing the LCVPs to clear the reefs with vital supplies. By nightfall, elements of Shoup's forces had fought their way across the airstrip and reached the south coast of the island. Shoup reported to General Smith on *Maryland*: "Casualties many; percentage not known; combat efficiency; we are winning."

By the morning of the third day on Tarawa, the Marines had clawed their way out of Red Beach 3 and were closing on the bombproof blockhouse which served as the headquarters of Rear-Admiral Shibasaki, commander of the island's defense forces. This was reduced to a smoke-blackened shambles as Marine assault engineers threw grenades down the blockhouse's air vents, followed by gasoline, to be detonated by TNT charges. That night the Japanese put in the long-expected counter-attack, but it was smashed up by naval gunfire as their troops assembled.

On 23 November, 3rd Battalion, 6th Marines, arrived on Betio, penning the remaining Japanese in the eastern end of the island while the weary survivors of the initial assault force reduced the lingering pockets of enemy resistance along Red Beach 1. The task was completed by the early afternoon, and at about the same time 3rd Battalion, 6th Marines, reached the eastern end of the island. At 1312 hrs General Smith declared Betio secure.

A small enemy force of about 175 men put up a final stand on the northern island of Buariki, but by 28 November the gruelling battle for Tarawa was over. Of the original Japanese garrison, only 146 were taken prisoner. Marine losses were 985 dead and 2,193 wounded, approximately 10 casualties per acre. Losses during the six-month campaign on Guadalcanal were just twice as heavy as those incurred in six days on

Tarawa. Nevertheless, the human cost of the operation was more than justified by the strategic results which flowed from it: the isolation of Rabaul, the move into the Marshalls and a drive through the Central Pacific as far west as the Palau Islands.

The savage struggle for Tarawa forced US planners to review their tactics for the assault of heavily defended islands. In spite of the ferocity of the naval barrage, the Japanese defenders had weathered the worst of the bombardment. In future operations the scale and duration of the "softening up" process would be increased and rockets and high-angle armor-piercing shells used to smash the enemy's heavily fortified bunkers and trench systems. Henceforward, Underwater Demolition Teams (UDTs) would precede landings to clear the way for improved LVTs. During the fighting on Tarawa, the Marines had been hampered by the unreliable radio links between the assault troops and the command ships lying offshore, which increased the difficulty of bringing in reinforcements to bolster a hard-pressed sector or of calling down artillery support on an enemy position. As the Marines fought their way inland, they were held up by a network of defenses which could not be silenced by relatively inaccurate naval gunfire. Short of heavy weapons, they took severe casualties when they closed on their well-dug-in enemy. The answer was provided by the development of "go-anywhere" amphibians, some armed with bunker-busting weapons. The American planners were confident that, with these improvements, an amphibious assault could get ashore, no matter what the obstacles.

On to the Marshalls

The campaign for the Marshalls — Operation Flintlock — began on 1 February 1944 when the newly-formed 4th Marine Division stormed ashore on the islands of Roi and Namur, the latter the site of a major Japanese air base on the northern tip of Kwajalein atoll. At the same time, 7th Infantry Division attacked Kwajalein Island, at the atoll's southern tip, while Majuro atoll — some 400 miles to the south-west — was secured by Army units for use as a fast carrier base.

In four days the central Solomons had been

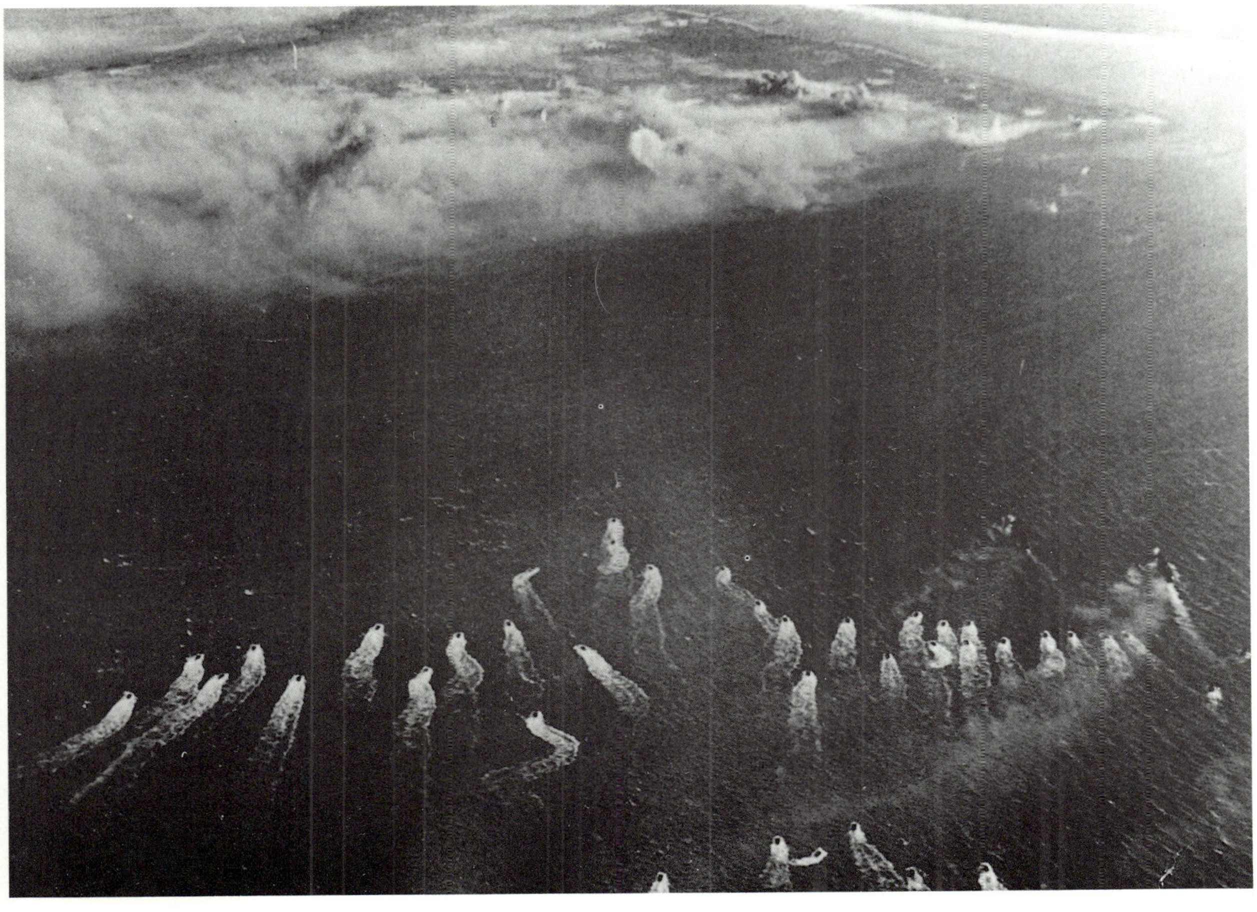

Below: Marine LVTS snake towards Engebi Island, in the Marshalls, on 18 February 1944. A dense pall of smoke raised by the naval and air bombardment hangs over the landing area. The LVT was one of the key elements in the Pacific campaign. LVTs were first used at Guadalcanal in August 1942, but only in a logistical role, ferrying ammunition and other supplies to the troops on the beaches. They were first in a tactical role employed at Tarawa in November 1943, clambering over the coral reefs which prevented deep-draughted vessels from reaching the shoreline. Their success at Tarawa resulted in the allocation of 300 amphibious craft to each Marine division. Inevitably, the LVTs suffered heavy losses, as they had to take the full brunt of the enemy's fire during the first phase of the assault. Further losses were caused by design faults. The LVT's bilge pumps were driven by the engine; if it failed, the LVT would quickly sink. Another fault lay in the organization of the amphibious assault units. Mother ships, tasked with delivering LVTs within range of their objective, often refused to repair or assist craft from other ships, condemning them to float off and eventually founder because of some minor defect. These initial shortcomings were ironed out and the LVT — of which 18,000 of various types were built between 1942 and 1945 — went on to play a vital role in the Pacific, enabling the troops to reach the shore relatively unscathed and providng them with their own heavy firepower.

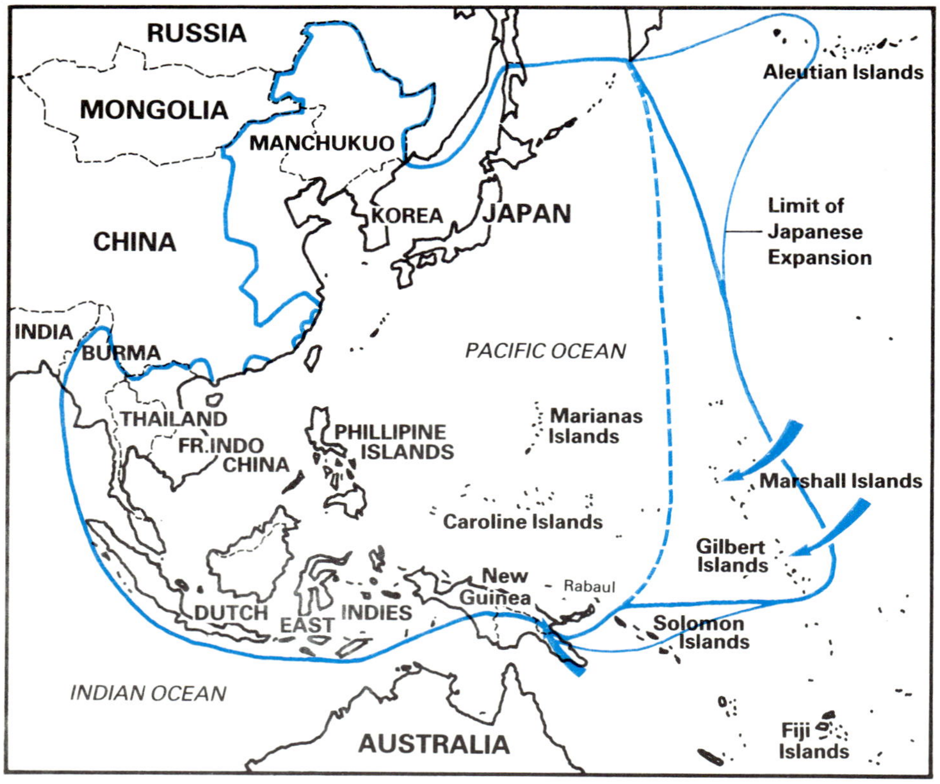

The United States' colossal economic and industrial power held the logistical key to victory in the Pacific. Above: Supplies pile up on the beach at Roi, in the Marshalls, in February 1944.

Left: Marines on the look-out for enemy snipers on Eniwetok atoll, in the Marshalls, February 1944. *Right:* A week after the landings on Green Island (15 February 1944), equipment and men move quickly off LSTs after a rough road has been cleared through the jungle by bulldozers.

Left: Seated in the back of a jeep during a tour of inspection in the Marshalls are Marine General Harry Schmidt (left) and Admirals Turner (center) and Connolly.

Below: Marines pause to salute the flag during the mopping-up operations on Roi. In the background are the smoldering remains of a three-storey enemy blockhouse knocked out after a furious bombardment.

taken at a cost of little more than 300 lives. The final target in this phase of the Pacific campaign was Eniwetok atoll, an important Japanese naval base 325 miles north-west of Kwajalein.

The threat of enemy naval or air intervention from the stronghold at Truk — 800 miles to the east — was eliminated by two days of bombing which destroyed 41 ships and 200 aircraft, turning Truk lagoon into an underwater graveyard

On Eniwetok, Marines of the 22nd Regiment and troops of 106th Infantry and 27th Division encountered stiff resistance when they went ashore on 17 February. In spite of a heavy air and naval bombardment, the atoll's defenders had to be winkled out yard by yard. The fighting was over by nightfall on 23 February, with not a single survivor from the atoll's garrison of 2,000 men. Overall Marine casualties in the Marshalls had been fewer than 600 killed.

The Battle for the Marianas

The next stepping stone in the slog across the Pacific was the Marianas chain, 1,000 miles north-west of Eniwetok, whose capture would provide bases from which naval and air forces could cut the lines of communication to the southern half of Japan's Pacific empire, and from which heavy and continuous bombing of the Japanese home islands could be made by the USAAF's new Boeing B-29 Superfortress long-range bombers.

As a preliminary to Operation Forager, Vice-Admiral Marc A. Mitscher's Task Force 58, comprising one battleship and four fast carrier task groups, secured total air supremacy over the Marianas. The islands of Saipan, Tinian and Guam formed the main Japanese defenses in the chain, garrisoned by troops of XXXI Army. For the amphibious operation, Admiral Nimitz assembled an assault force of 600 vessels and 127,000 men. The attack on the first objective, Saipan, inaugurated a new phase in the Pacific war. With the exception of Guadalcanal, all the American offensives in the Pacific had been directed against small islands, atolls or single small places on an extended coastline. The ensuing battles had been savage but short, with the Japanese dying to the last man. But the taking of Saipan, with its 72 miles of murderous terrain defended by 30,000 enemy troops, threatened a long campaign which, additionally, would have to be fought over a sizable civilian population.

At 0843 hrs on 15 June, after a prolonged bombardment, 2nd and 4th Marine Divisions (V Amphibious Corps) went ashore on Saipan's south-west coast. Within 30 minutes, 8,000 troops had been put ashore on a six-mile front. The plan was for 2nd Division to swing north to attack the 1,550ft Mount Tapotchau, the dominating feature on Saipan, while 4th Division thrust across the island to take Aslito airfield, five miles east of the beach-head.

By nightfall on 15 June, 20,000 troops had been landed on Saipan. In addition to both units landed in the first wave, the Americans had disembarked reserve units, seven pack howitzer battalions and elements of two armoured units.

Nevertheless, progress inland was slow. By sunset on 15 June, the assault troops had fought their way off the beaches, but only in the extreme south — the sector held by 25th Marines — had a line been consolidated along the first-day objective. Elsewhere the Marines had established only a shallow

Left: Dead Japanese soldiers litter the beach at Eniwetok.
Opposite, top: Seabees and Marines aboard an LVT at Roi, waiting for the order to land.
Opposite, bottom: Blood plasma is administered to a wounded Marine on Eniwetok.
Below, left: Sergeant Charles G. Bowers goes after a Japanese sniper with his Thompson sub-machine-gun on Eniwetok. The Marines had used the Thompson since the 1920s, and were probably the first military force to adopt the weapon, in Nicaragua in 1928. Although utterly reliable, the Thompson was expensive, slow to produce and in great demand by 1942. Its immediate replacement, the Reising 11.43-mm-caliber machine-gun, was not a success, rapidly clogging with dirt in combat, and was withdrawn after Guadalcanal. From 1943 the Marines adopted the M3 sub-machine gun, popularly known as the "Grease Gun", because of its resemblance to a mechanic's greasing tool.

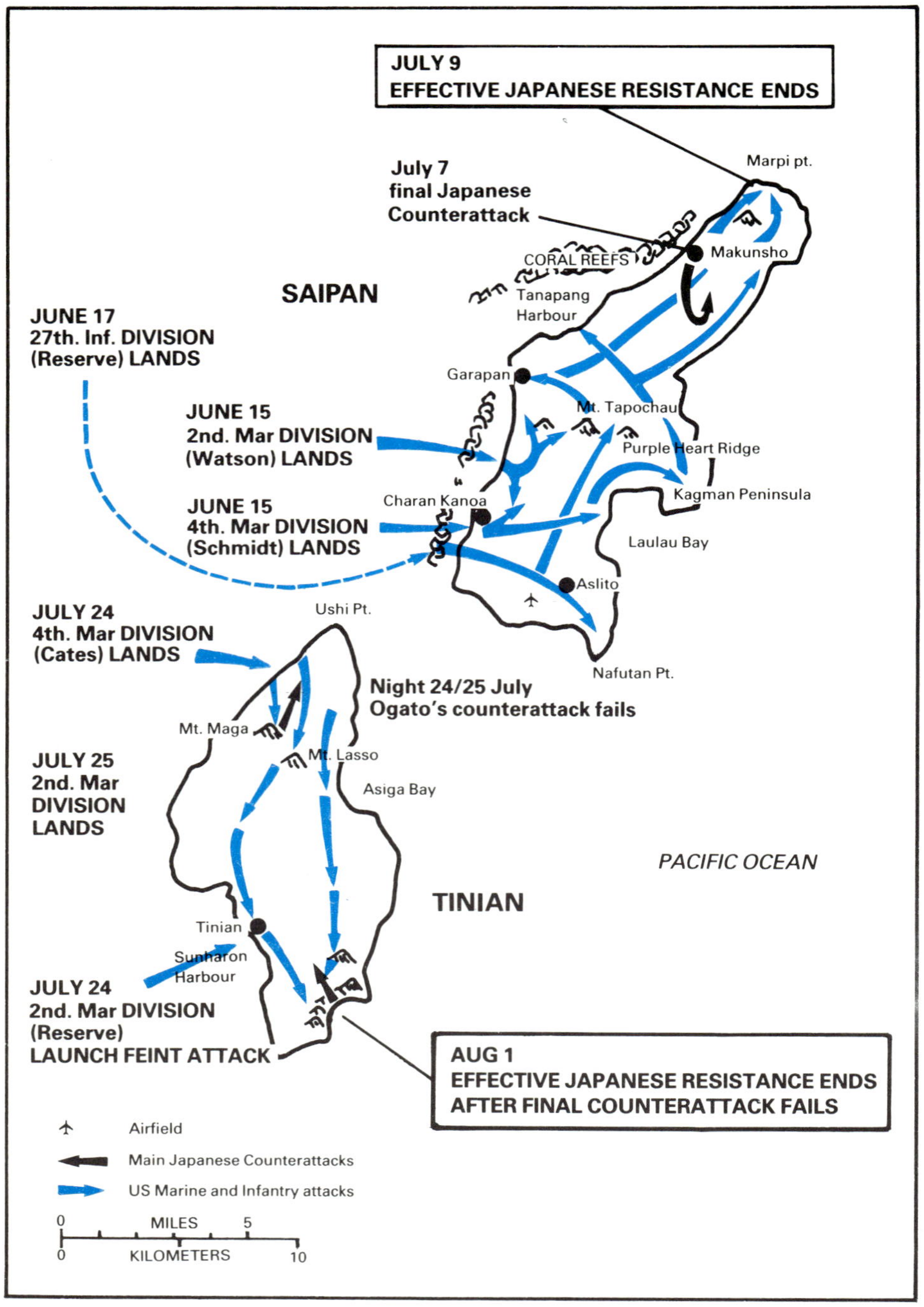

Above right: A Marine LVT(A), the "Buffalo", during the Saipan landings, 15 June 1944.
Right: Marines scramble for cover on the beach at Saipan shortly after H-Hour. In the background, smoke pours from a crippled "Buffalo".

defensive perimeter, and at the end of the day barely half of the designated beach-head had been secured. Enemy troops still clung to the ridges which overlooked the landing areas.

In gaining a foothold on Saipan, the Marines suffered about 2,000 dead and wounded, a casualty list which grew bigger that night when the enemy launched a counter-attack against 6th Marines, using Marines of their own and a reinforced armored battalion. But they had missed their chance to hit the Americans when they were at their most vulnerable — during the initial landings — and were thrown back by the Marines with the aid of naval gunfire, including that of the battleship *California.*

At this crucial point in the Marianas campaign, the Japanese suffered a crushing naval defeat. The American invasion fleet presented an irresistible target, and for its

Above: Which way to the front? Confusion on the beach at Saipan. In the background is a "Buffalo" mounting a 75-mm howitzer for heavy infantry support.

destruction the Japanese assembled a Mobile Fleet under Admiral Ozawa with nine aircraft carriers, five battleships, 13 cruisers, 28 destroyers and 478 aircraft. Opposing them was Admiral Mitscher's Task Force 58, with 15 carriers, seven battleships, 13 cruisers, 69 destroyers and 473 aircraft. Battle was joined on 19 June, when the first wave of Ozawa's carrier-borne aircraft was intercepted by 450 Grumman F6F Hellcat fighters. These powerful, chunky fighters were more than a match for their lightly armed adversaries and, in what became known as "The Marianas Turkey Shoot", over 250 Japanese aircraft were shot down at a cost of only 23 US 'planes. Meanwhile, American submarines had located Ozawa's Main Body, sinking the carriers *Taiho* (Ozawa's flagship) and *Shokaku*. Down to 100 aircraft, Ozawa headed north-west as darkness fell. Under the mistaken impression that he had inflicted heavy damage on the Task Force 58, he planned to renew the attack as soon as he had refueled and rearmed.

The Americans struck first. Late on the afternoon of 20 June, Mitscher found and attacked Ozawa's fleet, sinking the carrier *Hiyo* and two oilers as the Japanese withdrew towards Okinawa. Flying at the

limit of their range, many US aircraft were forced to ditch in the sea. In all, between 300 and 400 Japanese aircraft had been destroyed, with the disastrous attendant loss of aircrews. The carrier force survived, only to be annihilated later at the Battle of Leyte Gulf (24/25 October 1944).

On Saipan, the reserve 27th Infantry Division had been committed on 17 June. By nightfall on 22 June, the whole of southern Saipan had been cleared with the exception of the Nafutan headland. Thereafter, the commander of V Amphibious Corps, General Holland M. "Howlin' Mad" Smith, decided on an advance up the length of the island. Progress was slow, and the Army and Marine commanders fell to squabbling over what were essentially doctrinal differences. The Army believed in a cautious step-by-step approach. The Marines, aware of the vulnerability of the offshore fleet behind them, believed in taking ground at the utmost speed with scant regard for casualties. Holland Smith won the argument, and 27th Division's commander, Major-General Ralph Smith, was dismissed on 24 June.

His removal had little immediate effect in speeding up the breakthrough in "Death Valley", the Japanese defenses in the central part of Saipan. Mount Tapotchau fell to the 2nd Marine Division on 27 June, but it was not until the end of the month that 27th Infantry Division straightened the front by breaking through the center of the main line of Japanese resistance. By 6 July, the Japanese were penned in the northern third of Saipan. With defeat staring him in the face, Admiral Nagumo, Japanese naval commander on Saipan, committed suicide. His example was followed by the commander of the enemy Army forces, Lieutenant-General Sato, who slit open his belly with his sword and was then despatched by an officer with a pistol shot to the head. His final message to his troops was an invitation to follow him to glory against the enemy. They obeyed with a massed suicide charge in the Makunsha sector against two battalions of 27th Infantry, which lost about 1,000 men killed and wounded as they and the Marine gunners behind them were swamped by the banzai charge. The American reply was a concentrated artillery barrage which left some 3–4,000 Japanese corpses littered in heaps before their lines.

Saipan was secured on 9 July amid scenes of horror as the island's remaining troops and

22,000 civilians — driven mad by an irrational fear of being taken prisoner — performed an act of mass self-immolation. At one point, American troops watched helplessly as in the distance Japanese soldiers lined up to await decapitation by one of their officers. Lemming-like, whole families of civilians threw themselves off the 1000ft "suicide cliff" on to the jagged rocks below and from the bluffs at Marpi Point into the sea, deaf to frantic Japanese-language loudspeaker appeals assuring them that they would not be harmed. At Marpi, the dead and dying were clogged so thick that they fouled the propellers of the destroyers trying to pluck them from the sea. Of the Japanese military, the garrison of 32,000 men died almost to the last man. American casualties were almost as sobering — over 16,000 killed and wounded, of whom 12,934 were Marines.

Above: A Marine marksman takes aim at a Japanese sniper during the fighting on Saipan.

Top left: Marines in action with a captured Japanese mountain gun on Saipan.

Left: Waiting for the order to move out. The Marine in the foreground is carrying boxed mortar rounds lashed to his back.

After the fall of Saipan and the Battle of the Philippine Sea, the Japanese high command was forced to recognize that there was no longer any way of avoiding defeat. In the United States these victories laid the basis for President Roosevelt's nomination for an unprecedented fourth term. In Japan they led to the resignation, on 18 July, of the Tojo cabinet.

Tinian, three miles to the south of Saipan, was another tough proposition, its high, rocky coastline bristling with concealed batteries. The only chink in the coastal wall lay in front of the town of Tinian, on the south-west coast, but this had been heavily fortified by the Japanese. Accordingly, a landing was planned on the north-west tip of the island, where two undefended gaps in the jagged cliffs offered access to the interior. On 24 July an invasion fleet, spearheaded by 2nd Marine Division, made a demonstration in front of Tinian while 4th Marine Division landed in the north-west of the island. Achieving complete tactical surprise, they established a perimeter sufficiently firm to beat off a determined Japanese night attack. On 25 July, 2nd Marines went ashore, and within a week the island was in American hands.

The capture of Guam was of both strategic and symbolic importance. Not only did it contain three airstrips and a superb harbor, it had also fallen to the Japanese at the beginning of the Pacific war and would be the first piece of American territory to be liberated. On 21 July a combined assault on Guam was undertaken. 3rd Marine Division

Right: Japanese Zeros captured when the Marines overran Saipan's Aslito airfield.

landed on the west coast of the island four miles south of its capital, Agana; and 1st Provisional Marine Brigade and 77th Infantry Division came ashore to the south, at Agar, with the aim of trapping the Japanese forces in the Orote Peninsula, where the principal airfield was located.

The peninsula had been sealed off by 25 July, but that night the Japanese launched two frenzied counter-attacks, the first from the airfield and the second against 3rd Marines' beach-head. Both were driven back with horrific losses. Effective Japanese resistance on Guam had ended by 10 August 1944, after 10,000 of their troops had been killed. Mopping-up continued for a long time. The last Japanese to surrender on Guam did so in 1960.

A view of the beach at Pelelieu on D-Day, 15 September 1944. The landing area is shrouded with the smoke raised by a three-day naval and air bombardment.

Above: LVTs ferrying troops ashore at Pelelieu on D-Day.

Hell at Peleliu

With the capture of Guam, the Marianas were secured and the Carolines and Truk bypassed. The American high command now had their eyes fixed on the Philippines, and Admiral Nimitz decided that the next move was to be against the western Carolines and the Palau Islands. Lying 500 miles east of Mindanao in the southern Philippines, the Palau Islands stood athwart the left flank of the Central Pacific drive and threatened MacArthur's right flank as he moved up through the South-West Area towards the same goal. Nimitz insisted that the Palau Islands be taken rather than bypassed.

The main target was Peleliu, the principal island in the Palau group. Only two miles wide and six miles long, Peleliu's highest point is a hill named Umurbrogol Mountain, which was soon to earn the name "Bloody Nose Ridge". It was garrisoned by 10,000 Japanese troops dug deep into a honeycomb of interconnecting caves and bunkers — many of them fortified with sliding steel doors — lavishly stocked with food and ammunition. There were to be no mass suicide charges on Peleliu. The garrison's commander, Colonel Nakagawa, had been ordered to contain the Americans on their beach-head. If this failed, his troops were to withdraw to their prepared positions to enfilade the assault force with mortar and artillery fire. When overrun, they were to remain hidden and then strike at the enemy in the rear.

The Japanese defenses were subjected to a three-day air and naval bombardment before 1st Marine Division went ashore on the

south-west corner of Peleliu on 15 September. As they hit the beach, the Marines were met with a storm of fire from the island's defenders, who had sat out the barrage secure in their bunkers. Taking heavy casualties, the Marines fought their way across the island's airstrip, at which point Colonel Nakagawa ordered the withdrawal into the caves and strongpoints of Umurbrogol Mountain, by now stripped of all of its vegetation by gunfire and bombing.

On 17 September, the men of 1st Marine Regiment, commanded by Colonel Lewis B. "Chesty" Puller, began the nightmarish task of digging the Japanese out of their last-ditch defenses squad by squad, sometimes one by one. After three bloody days the 1st was replaced by 7th Marine Regiment, having suffered 1,700 casualties since landing on Peleliu.

At the same time, the Army's 81st Division was moved on to Peleliu, enabling 5th Marines to clear the northern sector of the island before turning to attack "Bloody Nose Ridge" from the rear. The ridge was surrounded by 321st Army Combat Team, 5th and 7th Marines, who then inched their way up its slopes, destroying Japanese

Opposite, top: Marines move up the beach at Pelelieu. They are wearing standard two-piece herringbone-twill fatigue suits and M1 helmets with "beach" camouflage cover.
Left: A Marine 4FU Corsair unleashes a rocket salvo against Japanese positions on Okinawaw, May-June 1945.

strongpoints with flamethrowers, grenades and satchel charges.

It took the Marines seven weeks to secure Peleliu and the surrounding islands in the group. The Japanese, who had succeeded in bringing in 4,000 reinforcements during the fighting, had lost 13,600 killed. Only 400 were taken prisoner. Each dead Japanese soldier had cost the Americans 1,600 rounds of light and heavy ordnance. They had also incurred 7,900 casualties (including 1,750 dead), of which over 6,000 were Marines.

Marine Aviation

In the New Georgia campaign, Marine air groups had developed highly effective close-support techniques under the direction of Major Wilfred Stiles. The basis of the system was the Air Liaison Party (ALP), operating in a radio-equipped jeep to call down strikes with maximum precision, thereby increasing hitting power and avoiding casualties among friendly troops.

The Marines were also the first to take delivery of the immensely powerful Vought 4FU Corsair carrier-borne fighter-bomber. Teething problems with the Corsair's carrier-landing ability led to all early deliveries going to the Marines, most notably to the flamboyant Major Gregory "Pappy" Boyington's "Black Sheep", VMF-214.

Combat with land-based Marine squadrons began in the Solomons in February 1943, and the Corsair quickly gained supremacy over the previously untroubled Japanese. Boyington was appointed commander of VMF-214 in September 1943 and led the unit until 3 January 1944, when he was shot down and became a prisoner of war. During his service with VMF-214, Boyington claimed 22 personal victories which, when added to his earlier six scores in China with the American Volunteer Group, made him the war's top-scoring Marine pilot, with 28 kills.

VMF-214's combat tour in the Solomons ended on 8 January, five days after Boyington was shot down over Rabaul. After carrier training in the United States, it briefly re-entered combat in March 1945 aboard the USS *Franklin*. The "Black Sheep" ended up as the seventh-highest-scoring Marine squadron of the war, with 127 victories, all but 30 of which were gained during Boyington's period of command.

The Marines achieved another first when, on 18 March 1944, VMF-111 exploited the Corsair's potential as a dive-bomber, attacking enemy anti-aircraft batteries with 1,000-lb bombs. Subsequently it was shown that the Corsair could be employed to devastating effect as a dive-bomber, delivering twice the payload of the specially-designed Dauntless in dives of up to 85 degrees. The fighter's six .5-in guns were used for strafing in the latter stages of the dive.

In the autumn of 1944 a vast amphibious armada carrying 200,000 men of Lieutenant-

Opposite, bottom: A Corsair of MAW-2 bombs enemy positions on Five Sisters Peak, Pelelieu, October 1944. Although designated a carrier-based fighter, the Corsair was rejected by the US Navy because its big nose restricted visibility and its undercarriage tended to make the aircraft bounce on touchdown. The Corsair entered service with the Marines in February 1943 in the South Pacific. Powered by a 2,000hp Pratt and Whitney engine, the Corsair had a maximum speed of 415mph at 20,000ft and a normal range of 1,000 miles. Variations of the F4U carried different armaments: for example, the F4U-1 had six 0.5-in Brownings while the F4U-1D fighter-bomber had provision for underwing bombs or rockets. In World War II, 2,140 enemy aircraft fell victim to the Corsair — a kill ratio of 11:1.

General Walter Kreuger's 6th Army sailed towards the eastern shore of Leyte in the Philippines. On 20 October, US 10th Corps and 24th Corps established a beach-head against light opposition from XVI Division of General Suzuki's XXXV Army. Japanese battleships and carriers attempted to draw off the US fleets from their support of the landings, but in a series of engagements between 23 and 25 October — the Battle of Leyte Gulf — the last vestiges of Japanese sea power were shattered.

Once the Americans were firmly established ashore, Marine air groups — many of them still equipped with the faithful Dauntless — were flown in to newly-liberated airfields to provide close support. Using the ground-attack techniques perfected during the Pacific war, these units gave the Army invaluable assistance in the protracted period of land fighting in the Philippines, and in the mopping-up operations which continued until the end of the war.

The Marine SBD squadrons swung into action during the Lingayen landings in January 1945, flying 255 sorties and dropping 104 tons of bombs for the loss of just one aircraft. During the capture of San Jose, north of Manila, 1st Infantry had suffered heavy casualties from strafing "friendly" USAAF fighters. Understandably, this made them extremely sceptical about the value of close support. However, on 18 February they required Marine dive-bomber assistance in a difficult operation. Nine SBDs made pinpoint attacks, placing all of their bombs within 30 yards of the target. Thereafter the SBDs remained on standby at the request of the Army.

Iwo Jima

While US 6th Army was fighting for control of Luzon, the northernmost island in the Philippines, V Amphibious Corps was preparing for the landings on the small volcanic island of Iwo Jima, and the bloodiest fight in the history of the Marine Corps.

Iwo Jima was a Japanese stronghold which could not be bypassed and left to "wither on the vine". Lying some 700 miles south of Tokyo, it formed an integral part of the inner ring of defenses around the Japanese home islands. From its two completed airfields, Japanese fighters could strike at the B-29 bomber streams flying to raid Japan's cities, and launch harassing raids on American forces on Saipan, Tinian and Guam. Less than two hours' flying time from Japan, Iwo Jima also provided an early-warning station from which to monitor US attacks on the home islands. The capture of Iwo Jima would provide the Americans with the bases from which short-range escort fighters could fly top cover for the B-29s and also act as an emergency stopping point for damaged bombers on their return journey.

Iwo Jima was administratively part of the prefecture of Tokyo, and its loss would deal a hammer blow to Japanese civilian morale. Recognizing its importance, and knowing that he could expect no help, the island's

Top: A Dauntless of VMSB-331 flying over Majuro atoll, in the Marshalls, in June 1944.
Above: A Marine Consolidated Vultee PB4Y-2 Privateer maritime patrol bomber in flight over the Pacific in April 1945. Derived from the B-24 Liberator, the Privateer saw postwar service in an electronic intelligence-gathering role.

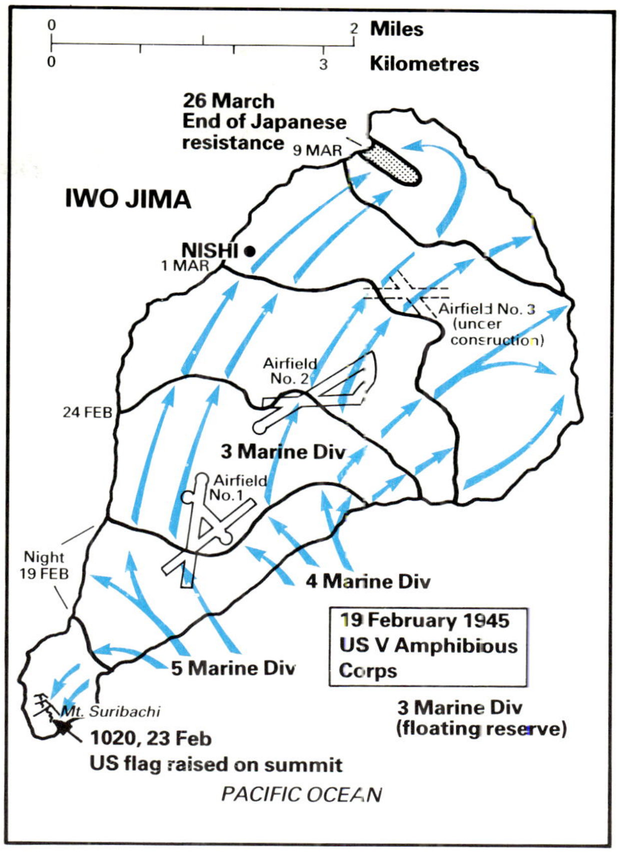

Below: Marines of 4th Division hug the black sand of Iwo Jima on D-Day, 19 February 1945. This was the division's fourth amphibious assault in 13 months.

commander, Lieutenant-General Kuribayashi, set about turning it into a death trap. Although only eight miles square, Iwo Jima was garrisoned by 21,000 men, packed with 1,500 pillboxes and blockhouses and criss-crossed with tunnels, trenches and minefields. The island's dominating natural feature, Mount Suribachi, a 600ft volcanic outcrop at the island's southern tip, was honeycombed with defensive positions.

Kuribayashi's artillery comprised 361 guns over 75-mm calibre, 300 anti-aircraft guns, 20,000 light guns, 130 howitzers, 12 heavy mortars, 60 anti-tank guns and 70 rocket launchers. Over 20 tanks were dug in, hull-down.

American preparations had been equally thorough. Preliminary planning for the operation had begun as early as September 1943. After the securing of the Marianas, General Holland M. "Howlin' Mad" Smith was placed in overall command of the assault on Iwo Jima. Spearheading the operation, under Major-General Harry Schmidt, was V Amphibious Corps, veterans of the Guadalcanal, Marshalls and Marianas campaigns.

Schmidt was given three divisions: the battle-hardened 3rd Marine Division, under Major-General G.B. Erskine, to be held offshore as a corps reserve; 4th Marine Division, under Major-General C.B. Cates; and 5th Division, under Major-General K.E. Rockey, untried as a formation in combat but containing a substantial percentage of veteran troops.

Having landed on Iwo Jima's south-eastern shore, under the glowering bulk of Mount Suribachi, the plan was for 4th and 5th Divisions to wheel north to assault the heavily fortified Motoyama Plateau, while 28th Marine Regiment undertook the daunting task of capturing Suribachi, from which the Japanese could observe every American movement.

On 8 December 1944, the Americans began the longest and heaviest aerial bombardment of the Pacific war, a 72-day "softening up" of Iwo Jima by B-24s and B-25s. Naval bombardment had begun in November and continued, with intervals, until 16 February 1945 when a 72-hour pre-assault barrage burst on Iwo Jima. Six battleships and their escorts pounded the Japanese positions, expending some 1,950 rounds of 16-in shell, 1,500 rounds of 14-in, 400 of 12-in, 1,700 of 8-in, 2,000 of 6-in and 31,000 of 5-in. Flying over the pitted, lunar landscape of Iwo Jima,

US aircraft napalmed and rocketed the Japanese defenses.

On 17 February, LCI gun- and rocket-boats came close inshore to cover the UDT teams clearing the beach approaches and checking the beach and surf conditions. Mistaking them for the first wave of the assault force, the Japanese opened up with their heaviest artillery, revealing their carefully-concealed positions.

D-Day, 19 February, dawned bright and clear, with a blue sky and calm sea. Offshore, 450 vessels of the US 5th Fleet maneuvered for the assault, while around them buzzed the 482 LVT(A)s which would carry the Marine battalions into action. As the Navy's ships laid down a creeping barrage — the first of the Pacific war — 68 LVT(A)s left their start line for the 4,000-yd dash to the shore. If all went according to schedule, the first seven battalions of Marines would be ashore within 45 minutes.

The Marines hit the thick black sand of Iwo Jima's beaches at 0902 hrs, 5th Division on the left, 4th on the right. Initially there was only light resistance, but when the Marines had pushed about 300 yard inlands they were suddenly raked with a hail of fire from concealed positions. Casualties were heavy, but the Japanese had made a fatal miscalculation. They had allowed the Marines to establish themselves ashore with all the equipment they needed to weather the storm. By 1030 hrs, elements of all eight assault battalions were ashore, and the bigger LSMs were bringing in tanks,

Above: Iwo Jima, 10 March 1945. Lieutenant-General Holland M. Smith (center) and Major-General Harry Schmidt in V Amphibious Corp's command post.
Above, left: Marines push inland on D+21, as the battle for Iwo Jima nears its end.

Left: Amid a litter of equipment, Marines entrench beneath the glowering bulk of Mount Suribachi.

Above: One of the indelible images of World War II — the Stars and Stripes are raised on Mount Suribachi.

bulldozers and artillery. By nightfall, nearly 30,000 Marines had been fed into the beachhead, securing the neck of the island, sealing off Mount Suribachi and gaining a lodgement at the corner of Airfield No. 1. But they had already sustained 2,400 casualties, including 600 dead.

It took 28th Regiment nearly three days to capture Mount Suribachi, an ominous bare mound resembling an ant-heap crawling with vicious soldier-ants. Fighting was frequently hand-to-hand. Private First Class Leo Jez was advancing towards a pillbox when a Japanese officer rushed him, wielding a sword. Jez deflected the blow, wrenched the sword from the officer's hand and, in a single movement, hacked off his head. The mountain was taken yard by yard, grenades, flamethrowers and satchel charges burning and blasting the Japanese in their dug-outs and bunkers.

The summit, now eerily silent, was reached at 1015 hrs on the 23rd by a platoon from 2nd Battalion led by Lieutenant Harold G. Schrier. They planted a small Stars and Stripes and were photographed by Sergeant Louis R. Lowery, a *Leatherneck* magazine photographer. The proceedings were interrupted by a brief firefight, during which Lowery was forced to roll some way down the mountain to avoid a grenade, smashing his camera in the process but saving the film. Two hours later, a second photographer, Associated Press man Joe Rosenthal, arrived at the summit, where the Marines were preparing to raise a second, bigger flag. It was one of the 18 pictures which Joe Rosenthal took during this flag-raising ceremony which became a symbol of the entire war for the United States.

There was much bitter fighting to come. On 24 February, 4th and 5th Divisions — reinforced by 3rd Division — confronted the main Japanese defensive line along the Motoyama Plateau. Their advance through a surreal landscape of jagged rocks and stinking sulfur pits was measured in tens rather than hundreds of yards; 28th Regiment, once again in the thick of the fighting, suffered 240 casualties daily for three days in a row. Artillery was useless against the Japanese positions and tanks were unable to maneuver in the broken terrain. Brilliantly camouflaged, the Japanese would often allow the Marines to pass through their positions before taking them in the rear with murderous bursts of fire.

On 28 February the assault began on the complex of trenches and bunkers on Hills 382 and 362A. On the following day, these objectives claimed the lives of three of the men who had helped to raise the flag on Mount Suribachi: Sergeant Hanson, Sergeant Strank and Corporal Block. Hill 382 fell on 1 March and Hill 362A was taken on the following day after a night attack.

It was not until D+18 (9 March) that patrols of 3rd Division broke through to the north-eastern shore of Iwo Jima. Another week elapsed before organized resistance collapsed, and mopping-up continued for another 10 days. On 4 March, 12 days before the island was declared secure, the first crippled B-29 flew in to land. On 7 April, 108 North American P-51 (Mustang) fighters took off from Iwo Jima to escort a daylight B-29 attack on Tokyo. Within three months over 850 B-29s had made emergency landings on the island. Without Iwo and its two airstrips, the majority would have been lost.

The price had been high. Of the 21,000 Japanese, only 200 were taken prisoner. Some 6,821 American soldiers and sailors lost their lives, of whom 5,931 were Marines. Over 18,000 were wounded, 17,372 from the Corps. The 28th Marines, in particular, had suffered grievous losses. Of the 3,900 men who had landed on Iwo Jima on 19 February, only 600 remained fit for action when the fighting ceased.

Left: D+18 on Iwo Jima: Marines wait warily as one of their comrades throws a hand grenade into an enemy emplacement.

The battle for Iwo Jima had been the only large-scale land action in the Pacific war in which American casualties outnumbered those of the Japanese. Its savagery prompted the pugnacious General Holland M. Smith to observe that it was the "toughest [fight] we've run across in 168 years". Admiral Nimitz declared that for the Marines on Iwo Jima, "uncommon valor was a common virtue", a tribute borne out by the awarding of 22 Congressional Medals of Honor — 12 of them posthumously — for acts of heroism during a single month of fighting. This tally represented more than one-quarter of the total awarded to Marines throughout the whole of World War Two. It was a chilling portent of what lay ahead if Japan itself was to be invaded.

Okinawa: the Last Battle

The final objective before the invasion of the home islands was Okinawa, the most important of the Ryuku Islands, almost exactly equidistant from Manila and Tokyo.

Okinawa was strategically located across the sea lanes to the south. Its capture would isolate Formosa and enable American medium bombers to strike at Japan. In enemy hands, it remained within range of land-based 'planes from Japan, but was out of range for similar American aircraft flying from the Philippines. Air cover for an invasion would have to be provided by carrier-based aircraft. In turn, the carriers would have to stay on station, making them highly vulnerable to kamikaze attacks.

A vast naval armada of more than 1,200 vessels was readied for the task of transporting 183,000 men of the US 10th Army (24th Army Corps, III Amphibious Corps) for the landings on Okinawa, codenamed Iceberg. The Marine element comprised 1st, 2nd and 6th Marine Divisions, plus a number of Marine air groups assigned to the covering Tactical Air Force.

On 1 April, after a week-long bombardment, the Americans went ashore on the south-west coast of the island, north of the city of Naha. Simultaneously, a feint was made against Okinawa's south-eastern shore by 2nd Marine Division. Initially, all went according to plan. Within three days the island had been split at the Ishikawa isthmus, with 6th Marine Division pushing northwards and the Army units advancing south. The Marines met only light resistance until they reached the Motobu Peninsula, which was secured by 20 April after a hard fight.

In the south, however, the Japanese had planned an unpleasant surprise for the Americans. General Ushijima was under orders to hold on until the kamikaze aircraft had sunk or driven off the American amphibious support ships. When this had been accomplished, he was to counter-attack and drive the invaders into the sea. Accordingly, he had fortified the southern third of Okinawa with a series of defensive lines, of which the most formidable began at Naha on the west coast, ran through the dominating high point of Shuri Castle and was anchored on the eastern shore at Yonabaru. It was Iwo Jima all over again, with the Japanese burrowed into caves and with reinforced emplacements on both forward and reverse slopes of their defensive lines.

While US 5th Fleet was assailed by waves of kamikaze aircraft, American soldiers and Marines battered away at Ushijima's lines. The first, the Machinato Line, held out for six days. Ushijima then fell back on his main line of defence — the Shuri Line — on which the Americans could make little or no impression. In desperation, the commander of 10th Army, General Simon Bolivar Buckner, considered an amphibious end-run, but all possible landing places were covered by heavy concentrations of Japanese artillery. There was nothing for it but to continue slogging away at the Shuri Line. The 6th Marine Division was brought down from the north and 1st Marine Division, until then

held in reserve, was thrown into action. The Marines of 6th Division took heavy casualties securing "Sugar Loaf Hill", and 1st Division was badly mauled in its fight for the Dakeshi Ridge and Wana Ridge. Under intensive bombardment, the Japanese fought every step of the way — during the campaign, American field artillery expended 1.7 million rounds. At sea the final toll of both kamikaze and conventional attacks was 36 US ships (most of them smaller types) sunk, 368 others damaged and 10,000 Navy personnel killed or wounded. But 5th Fleet still stood offshore.

The tide began to turn at the beginning of May. Unwisely, Ushijima yielded to demands from his subordinates for a major offensive, which was beaten back with the loss of 5,000 men, destroying his final reserves. On 21 May, he abandoned the Shuri Line and withdrew into last-ditch positions at the bottom of the island, across the Naha Plain, around the Yaeju Dake Escarpment. By the end of the month, Naha had been secured by the Marines after a savage house-to-house battle.

Not until mid-June did Marines and Army troops break the enemy positions on the escarpment. Okinawa's southern shore was reached on 19 June, and mopping-up operations continued for several more days. Three days before the battle ended, General Buckner was killed by an artillery shell. He was succeeded by Major-General Roy S. Geiger, a Marine aviator, who became the only Marine officer to have commanded a field army. Ushijima committed suicide,

Left: With a rifle lined on the entrance to a Japanese-held cave on Okinawa, Marines await the result of an explosive charge and prepare to shoot any enemy who attempt to escape. These bitterly contested cave positions formed the Japanese "Little Siegfried" line, defending the capital city of Naha.

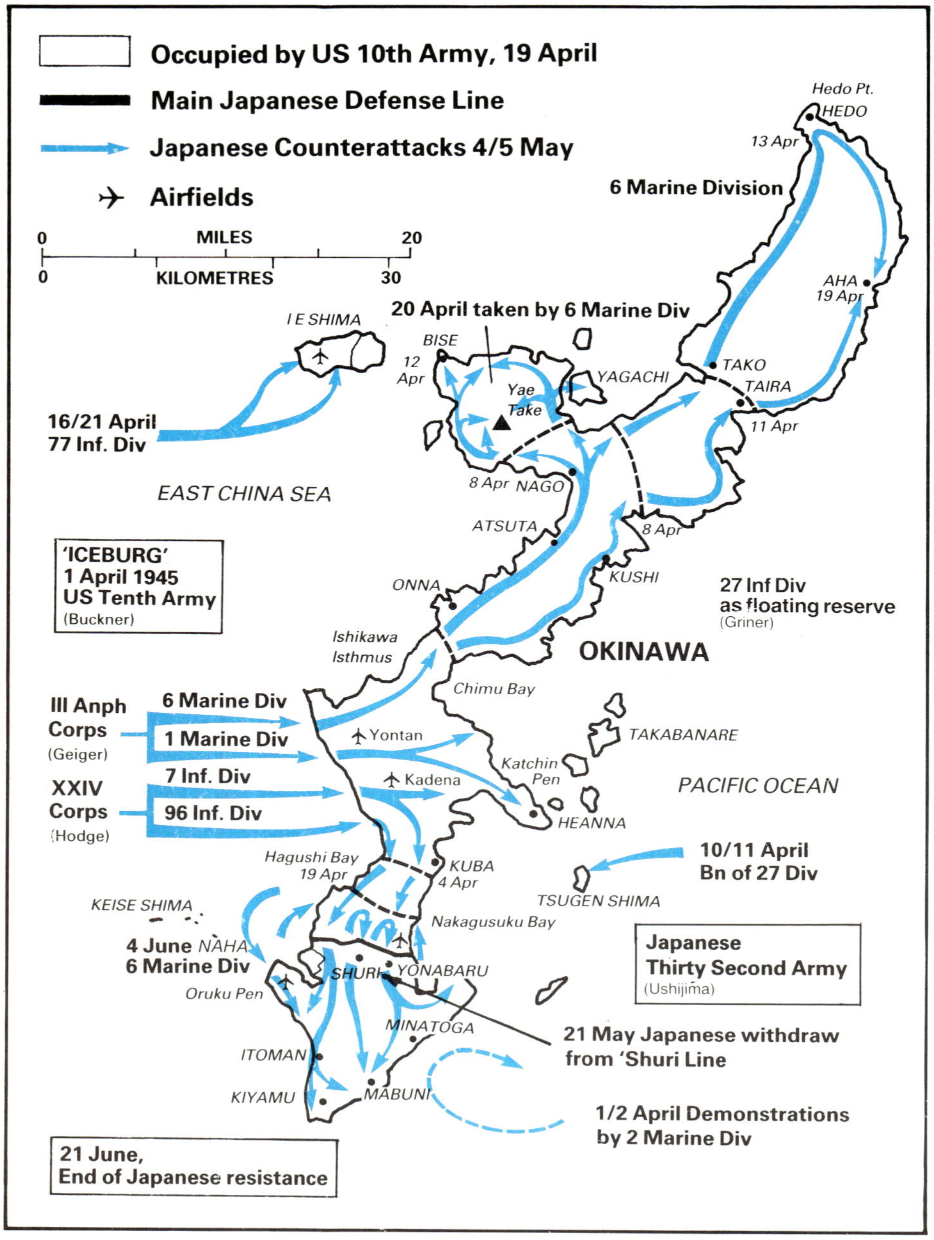

Right: Marines are pinned down by enemy sniper fire behind the gravestones of Okinawa's "Cemetery Ridge".
Opposite, top: Major-General Lemuel C. Shepherd, commander of 6th Marine Division, studies a map during the battle for Okinawa.
Opposite, bottom: A Marine Corps Grumman TBM Avenger drops two 500-lb bombs on enemy positions on Okinawa.

along with thousands of soldiers and civilians. Years after the war, one could still see grisly mounds of bleached bones at the bottom of the cliffs on Okinawa's southern shore.

Overall Japanese casualties in the campaign were 117,000, of whom 110,000 were killed. Marine Corps casualties were 19,000, including 3,000 dead. The complete 10th Army casualty list stood at 65,000 killed and wounded, another terrible reminder of the losses which would inevitably be incurred if the Japanese mounted an equally fanatical defense of the home islands.

President Truman decided to shorten the war, and at the same time send a pointed strategic message to the Soviet Union, by dropping the first offensive atomic bomb on Hiroshima, an important military headquarters and supply depot, on 6 August 1945. The bomb exploded over the center of the city, killing approximately 80,000 of its inhabitants. The Japanese high command still believed that continuing resistance might secure "honorable conditions". Accordingly, a second bomb was dropped on the seaport and industrial centre of Nagasaki on 9 August. Japan immediately surrendered unconditionally. A general cease-fire came into operation on 15 August and the official surrender was signed on 2 September aboard the battleship *Missouri* in Tokyo Bay. General MacArthur, resplendent in victory, declared "These proceedings are closed".

The Final Count

For the Marines, World War Two had been

overwhelmingly a Pacific War. Nearly 98 percent of all Marine officers served in this theatre, and 89 percent of enlisted men. It had been a unique campaign, fought in the steaming jungles of Guadalcanal, on the fire-swept beaches of Tarawa and the caves of Peleliu. The Corps' six divisions had made 15 landings. Its air arm, which by the war's end numbered 145 squadrons in 31 aircraft groups, scored 2,355 kills (top-scoring unit was VMF-121 with 208 victories). In the fighting from Wake to Okinawa, 19,733 Marines had died and 67,207 had been wounded in action. During the course of the war, 669,000 men had passed through its ranks. In August 1945 the Corps' strength stood at 458,000 officers and men. Between 1941 and 1945 it had met and overcome every challenge thrown at it by a ferociously determined enemy. Many new challenges lay ahead in the postwar world.

5:KEEPING THE PEACE:
The Postwar World 1946-1988

THE COLD WAR

In the immediate aftermath of the war, Marines of 4th, 2nd and 5th Marine Divisions formed part of the occupation forces in Japan. Other Marine units went to China to repatriate Japanese troops and to maintain the peace between Chiang Kai-shek's Nationalist forces and the Communists led by Mao Tse-tung. Marines also provided the garrisons in a number of Pacific islands.

Meanwhile, demobilization gathered pace. Within five months of the end of the war, 3rd, 4th and 5th Marine Divisions had been disbanded; 18 months later the strength of the Corps had fallen from a wartime peak of 485,000 to 92,000. Marine units were withdrawn from Japan in 1946, and the peacekeeping force in China was steadily reduced. The last Marines were withdrawn from China in May 1949. By the end of the year, the Communists had triumphed in China and Chiang Kai-shek's remaining forces had been driven from the mainland to the island of Formosa.

At home, the Corp's independence once more came under threat. The atom bomb appeared to be the harbinger of a revolution in the nature of warfare far more terrible than anything envisaged by Mitchell and Douhet in the 1920s. An amphibious force seemed an irrelevance in a world where the giant fist of air power cast its shadow. The United States Air Force, which was given independent status in 1947 by the National Security Act, was eager to swallow up Marine aviation. The Army maneuvered to absorb the remainder. But the National Security Act, which created a co-ordinated command of all the US armed forces under the Department of Defense, ensured that the Marine Corps retained its status as a separate military service within the US Department of the Navy. In January 1951, Congress passed the Douglas-Mansfield Bill, which formalized the Marines' separate identity and fixed the Corps' strength at four divisions and four air wings. The Commandant of the Marines was made a member of the US Joint Chiefs of Staff. By then, the Corps was involved in the largest military confrontation of the Cold War.

Left: On guard by dawn's early light.

Right: Two North Korean T-34 tanks destroyed by Marines in heavy fighting in the Naktong sector in August 1950.

Korea

Encircled by China, Japan and the Soviet Union, the mountainous and largely inhospitable peninsula of Korea had long been coveted by its powerful neighbours. In 1942 Korea was made an integral part of Japan. At the Cairo conference in December 1943, the Allies promised Korea its independence, a decision confirmed at the Potsdam Conference in July 1945. On 15 August 1945 the Allied Supreme Commander in the Far East, General Douglas MacArthur, directed all Japanese forces north of the 38th parallel in Korea to surrender to the Russians and those in the south to the Americans. Originally conceived as a temporary demarcation line, the 38th parallel quickly became a political frontier. In the South, the US-backed Republic of Korea (ROK) was established in August 1948; in the North, the Soviet Union set up the People's Democratic Republic of Korea (PDRK). By July 1949, all Soviet and US troops had been withdrawn from the peninsula; in their place was the experienced and well-equipped North Korean People's Army (NKPA) and the understrength and poorly-equipped ROK Army, armed with obsolescent weapons left behind by the Americans.

War broke out on 25 June 1950 when seven infantry divisions and a tank division of the NKPA attacked across the 38th parallel. Brushing aside the four ROK divisions in their path, the NKPA forces drove on the South Korean capital, Seoul. On the same day, the United Nations (UN) declared the invasion a "breach of the peace" and requested military aid for the ROK Army. On 27 June President Truman sent orders to MacArthur, now the US Commander Far East, to lend American air and naval assistance. Seoul fell on the 28th and two days later President Truman authorized the use of ground forces in Korea. American troops nearest to the battle zone were 7th, 24th and 25th Infantry Divisions, which made up part of Eighth Army's occupation forces in Japan. The US 24th Division, under General William Dean, began moving into Korea on 30 June. US troops engaged the NKPA for the first time on 5 July, near Osan, when a North Korean infantry division supported by tanks ran into "Task Force Smith", a makeshift American advance unit comprising a 105-mm howitzer battery, two rifle companies, two mortar platoons, six bazooka teams and a 75-mm recoilless rifle unit, under the command of Lieutenant-Colonel Charles B. Smith.

On 7 July, General MacArthur was appointed C-in-C UN Command. (British 27th Infantry Brigade arrived in Korea on 14 September 1950, and by the end of the war UN forces also included contingents from Turkey, Canada, Australia, Thailand, France, Greece, New Zealand, Holland, Colombia, Belgium, Ethiopia, Luxembourg and South Africa.) US 24th Division fought a gallant holding action against the NKPA — during which General Dean fell into the hands of the enemy — buying time until the arrival of two more US divisions, 25th Infantry and 1st Cavalry. By the beginning of August the ROK and US forces — now designated Eighth Army — had been pushed back into a confined area around Pusan on Korea's south-east coast, the so-called "Pusan Perimeter".

Opposite bottom: A marine takes up a firing position in a Korean schoolhouse.

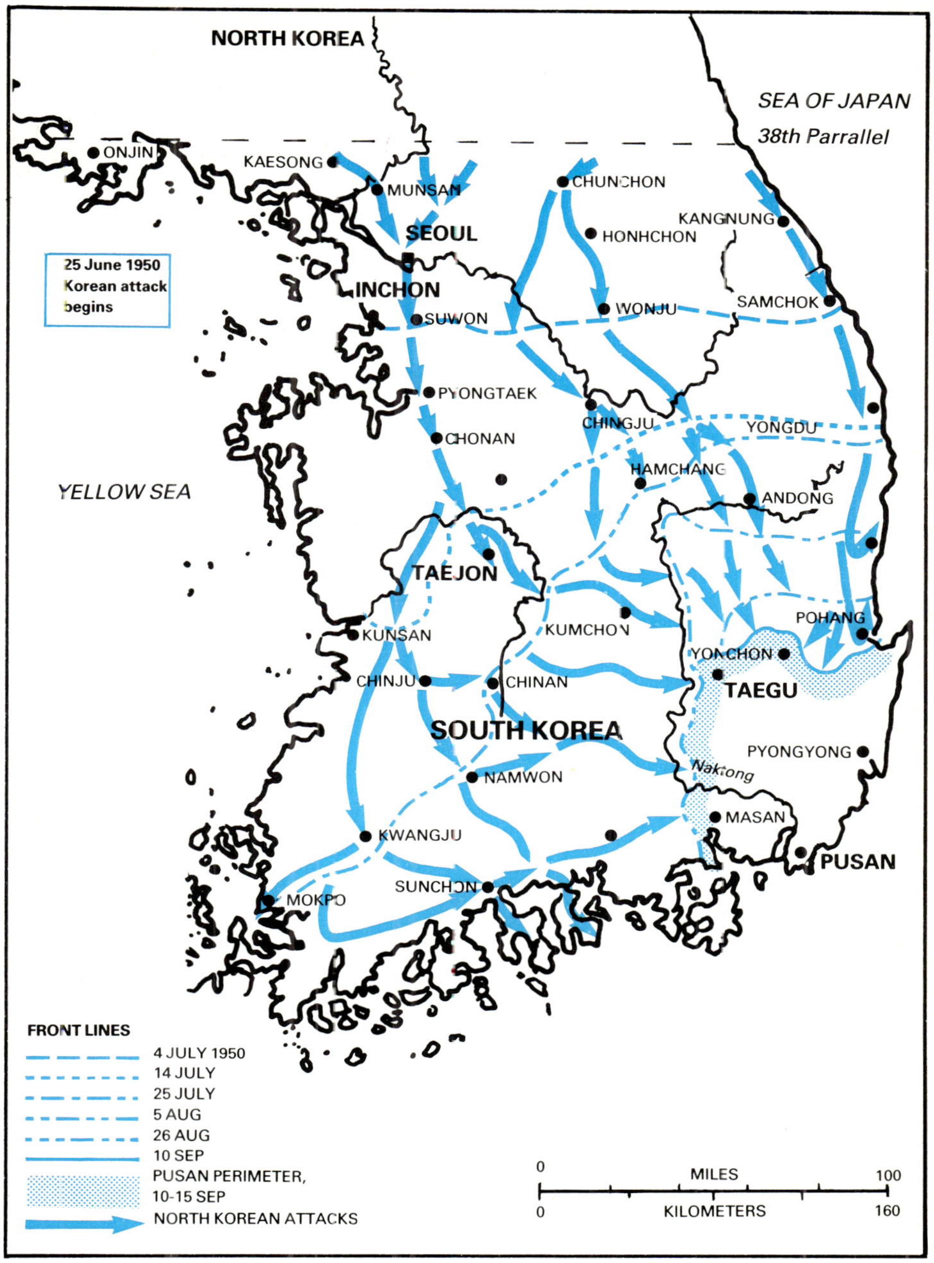
NORTH KOREA
SEA OF JAPAN
38th Parrallel
ONJIN
KAESONG
MUNSAN
CHUNCHON
HONHCHON
KANGNUNG
SEOUL
25 June 1950 Korean attack begins
INCHON
SUWON
WONJU
SAMCHOK
PYONGTAEK
CHINGJU
YONGDU
CHONAN
HAMCHANG
YELLOW SEA
ANDONG
TAEJON
POHANG
KUNSAN
KUMCHON
YONCHON
CHINJU
CHINAN
TAEGU
SOUTH KOREA
PYONGYONG
NAMWON
Naktong
MASAN
KWANGJU
PUSAN
SUNCHON
MOKPO
FRONT LINES
4 JULY 1950
14 JULY
25 JULY
5 AUG
26 AUG
10 SEP
PUSAN PERIMETER, 10-15 SEP
NORTH KOREAN ATTACKS
0
MILES
100
0
KILOMETERS
160

Opposite: Marine tank crewmen prepare ammunition for their M-26 Pershings aboard ship on their way to the Inchon beach-head, 13 September 1950.

Right: Marines of 5th regiment, 1st Marine Corps search two captured North Korean soldiers for concealed weapons.

As early as 2 July, General MacArthur had asked the US Joint Chiefs of Staff for a Marine Brigade. Within a week, 1st Provisional Marine Brigade had been activated at Camp Pendleton, taking most of the assets of the half-strength 1st Marine Division and 1st Marine Aircraft Wing. The core of the ground element was 5th Marine Regiment, commanded by Lieutenant-Colonel Raymond L. Murray, a veteran of Guadalcanal and Tarawa. The air elements came from Marine Aircraft Group 33 (MAG-33), including three squadrons of late-model F-4U Corsair fighter-bombers. The Brigade's 6,534 men sailed from San Diego on 12 July, arriving in Pusan on 2 August.

The Marines' immediate task was to plug holes in the buckling perimeter as they

Right: A North Korean soldier, captured in the Naktong sector, proves a reluctant subject for the camera. Around his neck hangs a prisoner-of-war tag.

opened. Their Brigade commander, General Edward A. Craig, told them: "It will be costly fighting against a numerically superior enemy . . . Marines have never lost a battle. This Brigade will not be the first to establish such a precedent".

The 1st Marine Brigade entered action on 7 August as part of Task Force Kean (1st Marines, an ROK Army battalion and 5th and 35th Army Regimental Combat teams), named after the commander of 25th Infantry Division, Major-General William Kean. Having blunted a drive on the Pusan Perimeter from the south-west by the NKPA 6th Division, Task Force Kean launched a counter-offensive against Chinju, enemy headquarters for the south-western sector. By nightfall on 11 August, 1st Marine Brigade was dug in above Changchon, 26 miles deep into enemy territory.

Shortly afterwards, the Marines were withdrawn to shore up another hard-pressed stretch of the perimeter along the southern section of the Naktong River, fighting alongside 24th Infantry Division. MAG-33 was also in action: two fighter-bomber squadrons were flying close-support missions from the decks of carriers of US Seventh Fleet; a night-fighter squadron (VMF (N)-513) was flying missions from Japan; and an observation squadron (VMO-6), equipped with helicopters and light aircraft, supported ground elements in the Pusan perimeter.

Fighting in the Naktong sector became critical in mid-August as the NKPA 4th Division drove a salient across the river, threatening the provincial capital of Taegu, at the north-western tip of the UN enclave. From its position on the right of US 24th Division, 1st Marine Brigade drove the North Koreans from their dominating position of "No-Name Ridge" and back across the Naktong River.

With the Pusan perimeter temporarily secured, the Marines were withdrawn to act as the reserve for 25th Infantry Division and to rest and replenish their equipment. On 31 August the NKPA renewed its offensive along the Naktong, breaking the line in many places. The 1st Marine Brigade was rushed into the front line, flanking 2nd Infantry Division in the Yongsan–Changnyong sector. After three days of hard fighting, the Marines mauled the NKPA 9th Division, retrieving many pieces of abandoned American ordnance and leaving the countryside "littered with enough North Korean arms, tanks and vehicles to equip a small army". In the space of six weeks the Marines had destroyed two crack North Korean divisions.

On 5 September the Brigade was ordered back to a staging area in Pusan "for further operations against the enemy". A week later the Brigade was absorbed by the newly-arrived 1st Marine Division.

From the outset, General MacArthur had determined on reversing the desperate situation in Korea with an amphibious assault in the enemy's rear. On 25 July he secured permission from the Joint Chiefs of Staff to bring the "skeletonized" 1st Marine Division up to wartime strength in readiness for an amphibious operation in Korea. This required a supreme effort of organization. The activation of the Provisional Brigade had reduced 1st Marine Division to 3,386 officers and men, less than half the strength of a single regiment.

Within 53 days its strength had been increased to 17,162 men. The understrength 2nd Marine Division had provided 7,812 men; 3,630 Marines had been pulled in from posts and stations throughout the world; and 2,891 combat-ready Reservists, most of them World War Two veterans, had been provided by a mobilization of the entire Organized Reserve. A second infantry regiment, 1st Marines, was assembled in ten days by Colonel Lewis B. "Chesty" Puller, who had commanded the same regiment at Cape Gloucester and Peleliu. The 1st Marine Division would become the spearhead of the newly activated X Corps, which would also include US 7th Infantry Division, reinforced with 8,000 ROK troops.

Inchon

X Corps, commanded by Major-General Edward M. Almond, was to deliver the first major UN counterstroke at Inchon, on Korea's western coast, 150 miles north of Pusan and about 20 miles west of Seoul.

The amphibious landings at Inchon were MacArthur's last great strategic stroke, both daring in conception and posing formidable problems of execution. In the headlong postwar demobilization, most of the US Navy's amphibious shipping had been given away or allowed to fall into disrepair. Ironically, 30 of the 47 LSTs required for the operation had to be retrieved complete with their crews from the Japanese, who were using them as inter-island ferries.

Intelligence reports indicated that there were about 21,000 enemy troops in the immediate area of Inchon — most of them of poor quality — with about 2,000 stationed in the town itself. These were less problematic than the hydrographic idiosyncracies of the target area. Inchon's harbor — guarded by the shore batteries on Wolmi-do Island — could be reached only through the narrow, winding Flying Fish Channel, which was subject to daily tidal variations of 30ft. On 15 September, the day scheduled for the landings (Operation Chromite), morning tide would be at 0659 hrs and evening tide at 1919 hrs. When the tide fell, the currents ripped out of the channel at up to eight knots, exposing glutinous mud flats on which even the assault force's amphibious tractors would stick fast. Pinpoint navigation would be essential for the success of the operation; it was the typhoon season; and there would be no time for rehearsals.

The assault force for the landing was Amphibious Group One, with 1st Marine Division embarked, under the US Navy's amphibious expert, Rear-Admiral James H. Doyle. The 7th Marines, the Division's third infantry regiment, were still *en route* to Korea and would not be there in time for the operation. A reserve regiment was provided by the newly-formed 1st Korean Marine Corps (KMC) Regiment along with its Marine Corps advisers.

The 3rd Battalion, 5th Marines, was to seize Wolmi-do — which was connected to the Inchon docks by a 600-yd causeway — on the

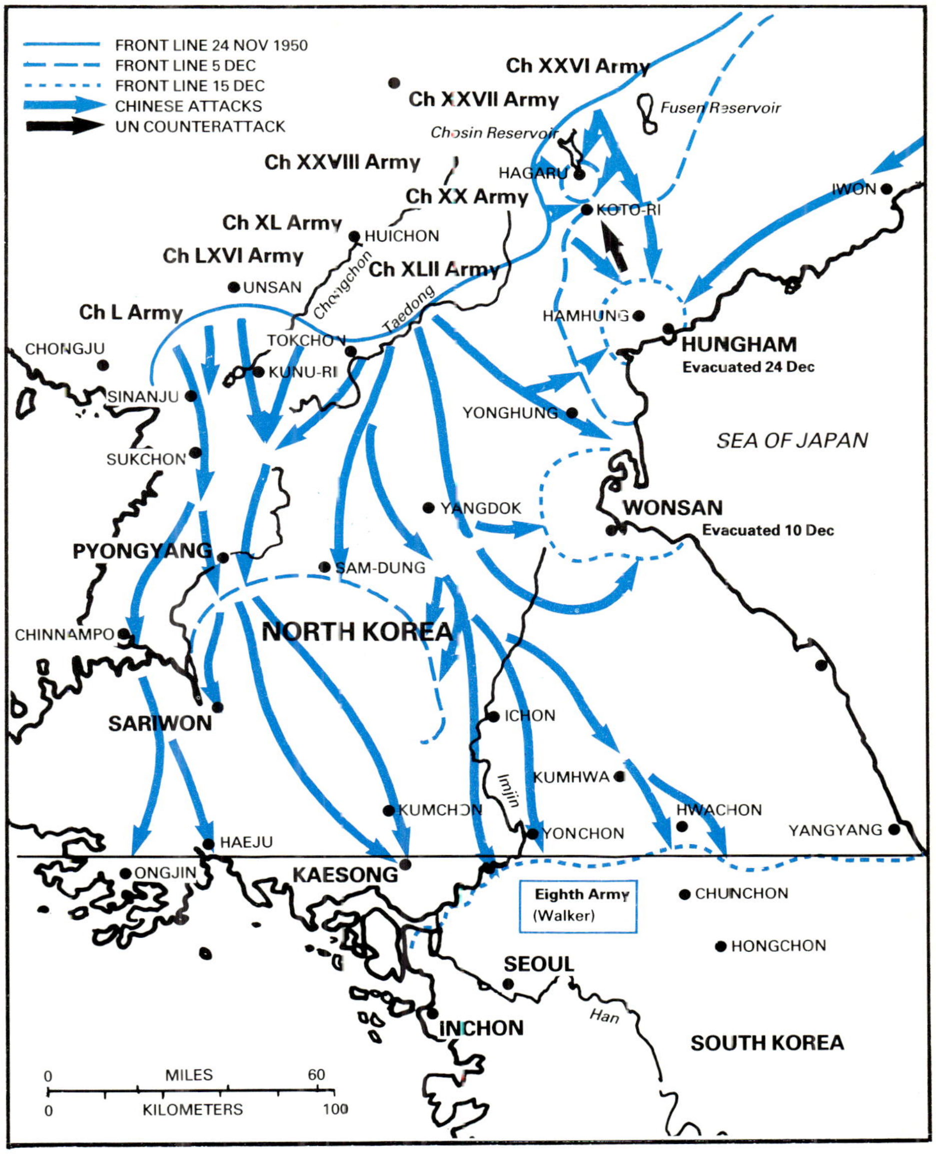

Left: Smoke raised by naval gunfire covers the landing areas at Inchon. Geysers of water can be seen near Inchon's docks, in the center of the picture.

Right: Marines use scaling ladders during the amphibious assault at Inchon, 15 September 1950. Only on Blue Beach, where 1st Marines landed, were there any problems. Some landing craft ran aground on mud flats 500yds from the shore and part of the reserve battalion, confused by the smoke-shrouded landing area, came ashore in the wrong place.
Far right: A tank-infantry team of 1st Marine Division pushes inland after the Inchon landings.
Below: Marines and an M-26 Pershing tank move through the bullet-swept outskirts of Seoul on 25 September. In Korea, 1st Marine Division's three infantry regiments and single artillery regiment (11th Marines) were supported by a number of organic and attached battalions, including: 1st Amphibian Tractor, 1st Armored Amphibian, 1st Combat Service, 1st Engineer, 1st Medical, 1st Motor Transport, 7th Motor Transport, 1st Ordnance, 1st Service, 1st Shore Party, 1st Signal, 1st Tank (with M-26 Pershings), the Headquarters Battalion, and 1st Combat Service Group. Attached to the Division were two aviation units — Marine Observation Squadron 6 and Marine Helicopter Transport Squadron 161.

morning tide across Green Beach. There would then be a 12-hour delay until the evening tide, when the remainder of 5th Marines would land on Red Beach, to the north of Wolmi-do, while 1st Marines came ashore at Blue Beach in the inner harbor. These landing areas were not "beaches" in the strict sense of the term. The harbor had granite walls at the sea's edge, which would have to be scaled.

Tasked with close air support was 1st Marine Aircraft Wing, flying from the decks of Seventh Fleet's light carriers USS *Sicily* and *Baodeng Strait* (dubbed the "Bing Ding" by the Marines). The "softening-up" of Wolmi-do began on 10 September with napalm strikes flown by Marine squadrons VMF-214 and VMF-323. For two days before the landing, US and British cruisers and destroyers pounded Wolmi-do and the Inchon waterfront.

Before daylight on 15 September, the Attack Force moved up the channel towards Inchon, with General MacArthur aboard the command ship *McKinley*. The shore bombardment lifted at 0545 hrs and within 15 minutes the men of 3rd Battalion, 5th Marines, were aboard their landing craft. The air attacks lifted at 0615 hrs and were followed by a blanket salvo of 5-in rockets from three rocket boats. The first wave of Marines went ashore on Green Beach at 0633 hrs, meeting only light resistance. By noon the battalion had secured the island and eliminated the enemy positions flanking the approaches to the main landing area at Inchon.

Naval gunfire and air strikes continued to plaster Inchon, and at 1733 hrs leading elements of 1st and 2nd Battalions, 5th Marines, scaled the sea wall at Red Beach and began to push south and south-east. On the right flank, another naval bombardment preceded the first wave of Puller's 1st Marines, aiming a swinging right hook to cut off Inchon from Seoul.

By the afternoon of D+1, the force beach-head line had been secured. The assault phase of the operation, launched under the most difficult hydrographic conditions, had been a stunning success. Marine casualties were 20 killed in action, one dead from wounds, one missing in action and 174 wounded. On the same day, 1st and 5th Marines began to push down the road to Seoul, leaving the Korean Marines to mop up resistance in Inchon. On the 17th, 5th Marines secured Kimpo airfield, north-east of Inchon, and soon US C-54 and C-119 transports were landing every few minutes under the direction of Major-General William H. Tunner, mastermind of the 1948 Berlin Airlift. On the 18th, 7th Infantry Division began to come ashore at Inchon, moving up on the Marines' right flank. Two days later 3rd Battalion, 5th Marines, crossed the Han River in amphibious tractors. Enemy resistance was now stiffening, but by 25 September the Marines had fought their way through to the outskirts of Seoul. On the 26th an armored breakout from the Pusan perimeter joined hands with the Inchon landing force. The battle for Seoul lasted for two days before NKPA resistance collapsed. On the 29th South Korea's President Syngman Rhee, escorted by General MacArthur, entered Seoul, three months and four days after the NKPA had launched its invasion of South Korea.

Retreat from Chosin

On 27 September General MacArthur was ordered by President Truman to "complete the destruction of the North Korean forces", with authority to cross the 38th parallel into North Korea. At the beginning of October, the UN General Assembly passed a resolution calling for the "establishment of a united, independent and democratic government in the sovereign state of Korea". The roles had been reversed; now the UN was attempting to unite the Korean peninsula by force of arms. The North Korean capital, Pyongyang, fell on 20 October. By the third week in November, UN forces had occupied two-thirds of North Korea, pressing to within a few miles of the Yalu River, which marked the border with China.

On 26 October, 1st Marine Division had been landed at Wonsan, on Korea's east coast, where it formed part of X Corps' three-pronged advance on the Yalu River, which forms the border between North Korea and Manchuria. ROK I Corps was assigned to the right flank; US 7th Infantry Division was in the center; and 1st Marines took up the left flank; corps reserve was provided by US 3rd Infantry Division.

The only route to the Marines' objective was over a narrow road through wild mountain terrain which rose to a height of 4,000ft. Winter was closing in, bringing sub-zero temperatures and marrow-chilling winds. The Marines began their advance to the Yalu at Hamhung, 70 miles north of Wonsan. Their sector of responsibility — approximately 200 miles long and 40 miles wide — would have taxed the abilities of a corps-sized formation. The Marines' commander, Major-General Oliver P. Smith, was not only worried about his division's widely dispersed positions. He was also concerned about the 80-mile gap which separated X Corps' left flank from Eighth Army, advancing in the west but cut off from 1st Marines by Korea's mountainous spine. Another cause for alarm came with intelligence reports that elements of 12 Chinese divisions had been identified in the Chosin reservoir area, 70 miles to the north.

At the beginning of October, the UN high command had been confident that their troops would be "home by Christmas", an unfortunate echo of the mood in August 1914 and as equally ill-founded. As he pushed northwards, the wary General Smith left behind stockpiles of arms, ammunition, fuel and supplies, to minimize 1st Marines' vulnerability if they were obliged to withdraw through the granite gullies and gorges of the road which led to the Chosin reservoir.

On 4 November, 7th Marines' Regimental Combat Team reached Chinhung-ni, where they encountered Chinese Communist forces (CCF) dug-in in strength. In a five-day battle the Marines so severely mauled major

Right: An impressive display of logistics at Wonsan as tank landing ships disgorge their cargo, M-24 Chaffees.

Below: Wave after wave of DUKWs bring 1st Marine Division ashore at Wonsan, on the eastern coast of North Korea, 25 October 1950. The landings were unopposed as the area had already been secured by ROK forces.

elements of 124th CCF Division that it was never effective again as an organic unit.

By 25 November, 5th and 7th Marines were on the western side of the Chosin reservoir, preparing to launch a westward thrust out of its mountain valley. Fourteen miles to the south-east, across a frozen, windswept plateau, the forward elements of the divisional command post and a battalion of 1st Marines held Hagaru-ri at the southern tip of the reservoir. Further south were the remaining two battalions of 1st Marines, dug in on the icy slopes of Koto-ri. Behind them lay the Funchilin Pass, gateway to the high plateau of the Tobaksan Mountain range which cradles the Chosin reservoir.

On that day the Korean War entered its second phase. The NKPA, reinforced by 200,000 CCF troops acting as "volunteers", launched a massive counter-offensive across the Yalu. The main blow fell on the right flank of Eighth Army, which by the end of the year had been bundled back to the 38th parallel. On 26 November, 1st Marine Division had been ordered to attack westwards to sever the enemy lines of communication at Mupyong-ri. But before this operation could gain any momentum, the Chinese delivered their second blow, this time aimed at X Corps.

On the night of 27 November, the assault battlions of the CCF North Army Group hurled themselves against the 5th and 7th Marines' perimeter at Yudam-ni, 1st Marines at Koto-ri and the three-battalion task force of US 7th Infantry Division east of the Chosin reservoir. While the Chinese 79th and 89th Divisions battered at the Marines' positions at Yudam-ni, 59th Division cut the 14-mile section of the road back to Hagaru. On the next day the Chinese attacked Hagaru, with its vital fuel dumps and nearly-completed airstrip, opening a gap in its four-mile perimeter at East Hill, a position dominating the town.

The 1st Marine Division was now encircled by Communist forces, and General Almond authorized General Smith to abandon his heavy equipment to speed a breakout and withdrawal. Smith declined to leave his heavy equipment behind him, adding that he would withdraw only as rapidly as he could evacuate his wounded.

On the 29th, operations began to open the road to Hagaru. Six days later, 5th and 7th Marines fought their way through, many of them marching in cadence and singing the Marines' hymn as they arrived. With them

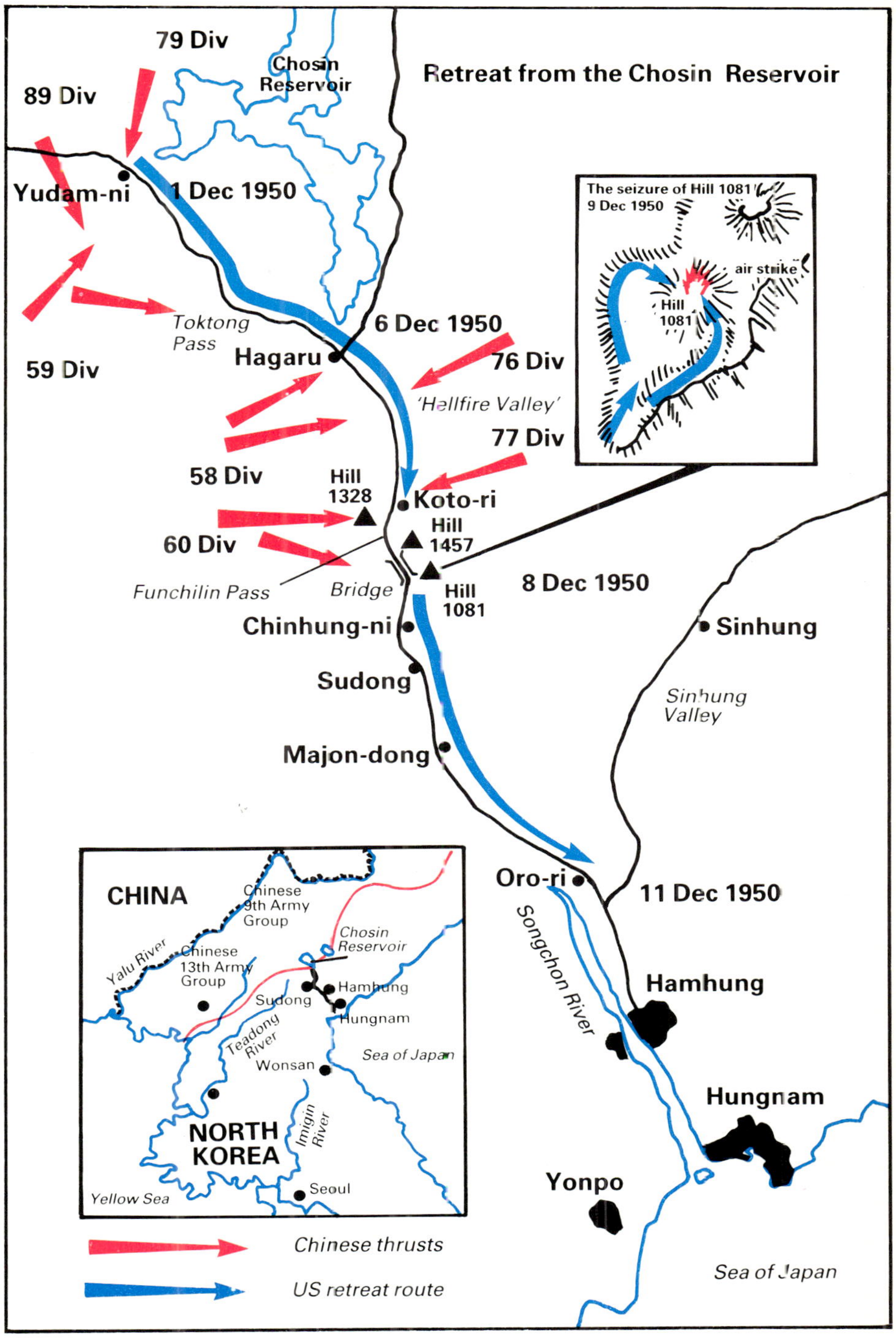

they brought 1,500 casualties. Hagaru's airstrip — hacked out of frozen ground in 12 days — had become operational on 1 December. In five days 3,150 Marines and 1,137 soldiers were evacuated by Marine R4Ds and Air Force C-47s. The same aircraft flew in 537 replacements, some of them recently recovered from the wounds they had received in the fighting for Inchon and Seoul.

Once the Marines were in the Hagaru perimeter, General Smith ordered the division to break out to Hamhung, 56 miles to the south-east. When asked by a newsman if this was a retreat, Smith explained that "retreat" was not the correct term: "Retreat, hell! We're just attacking in another direction!"

On 6 December, 7th Marines struck out for Koto-ri while 5th Marines plugged the gap in East Hill. The two-day breakout to Koto-ri was achieved with the loss of 103 dead, 506 wounded and seven missing in action. Close air support was provided by land- and carrier-based aircraft. Smith had now concentrated 14,000 men — including all three of his infantry regiments and some British commandos — but was keenly aware that the Chinese would reserve their main effort for the ten-mile gauntlet down the Funchilin Pass to Chinhung-ni.

At a hairpin bend, where the road clung to the southern side of Hill 1081, the enemy had blown up a narrow concrete bridge. To save his heavy equipment, particularly his tanks, Smith decided to replace the bridge with a 16-ton steel bridge, air-dropped in sections, an operation never attempted before. On 9 December the site on Hill 1081 was secured by 1st Battalion, 1st Marines. The steel sections were flown in from the emergency airstrip at Koto-ri and assembled in three hours under heavy enemy fire. Throughout the operation there had been fierce fighting along the length of the road, from the rearguard action at Koto-ri to the advanced elements pushing into the outskirts of Chinhung-ni. There, they linked with Brigadier-General Armistead Mead's Task Force Dog, which had fought its way up from Hamhung to clear a path for 1st Marine Division on the last leg of its epic journey. By nightfall on 11 December, all of 1st Marine

Below: Corsair fighter-bombers of the four squadrons of MAG-12 at Yonpo airfield, December 1950. They provided crucial close support during 1st Marines' epic withdrawal from Chosin. The very heavy AU-1 attack Corsair saw much action in Korea, carrying a 4,000-lb load at speeds which seldom exceeded 240mph.

Above: In heavy snow, 1st Marine Division readies itself for the break-out from Koto-ri. Only about 25 percent of the division had been issued with cold-weather gear — mountain sleeping bags, parkas, wind-proof trousers, shoe-pacs and heavy woolen socks. The parkas were Navy-type, heavy and long cut, better suited to watch-standing than long marches through freezing mountain terrain in which the temperatures plummeted to 20 degrees Fahrenheit below zero with a 20-30 knot wind.

Division's units were safely in their assembly areas in the Hamhung–Hungnam sector.

General Almond had been ordered to redeploy to the Pusan area, and the evacuation of the Hungnam perimeter was completed by 24 December. The 1st Marine Division sailed on 15 December. Since landing at Wonsan, the Marines had suffered 4,418 battle casualties — 718 dead, 192 missing and 3,508 wounded. There were also 7,313 non-combat casualties, most of them suffering the effects of the extreme cold. They had fought the 20th, 26th, 27th and 42nd CCF Armies, 14 divisions in all, inflicting an estimated 40,000 casualties on the enemy. *Time* magazine declared that the Marines' fighting retreat was: "A battle unparalleled in US military history. It has some of the aspects of Bataan, some of Anzio, some of Dunkirk, some of Valley Forge." In less florid but no less telling military language, the 1st Marines Division had achieved a signal tactical triumph by "coming through with operable equipment, with wounded properly evacuated, and with tactical integrity".

A War of Attrition

No sooner had General Walton Walker stabilized Eighth Army's front on the 38th parallel than the Chinese launched a second offensive, occupying Seoul on 4 January. On 23 December, General Walker was killed in a road accident and replaced by Lieutenant-General Matthew B. Ridgway, who assumed control of all land operations under General MacArthur.

The Marines were now drawn into a savage war of attrition. There were to be no more amphibious landings in Korea. Instead, the Marines were assigned missions similar to those undertaken by Army divisions, and its air elements were placed under the control of the USAF.

After the evacuation of Hungnam, 1st Marine Division had been placed in the Eighth Army reserve to re-equip and receive replenishment. The launching of the renewed Communist offensive brought them back to the front line. On 10 January 1951, Chinese forces broke through the UN lines on the right flank of US 2nd Division, infiltrating ROK III Corps. The Marines were immediately rushed northward from Masan to the Andong–Yondok road, where enemy forces were threatening to cut the supply lines of the ROK troops in the eastern sector.

The Marines stabilized this front and were then assigned the task of destroying enemy guerrilla forces — the remnants of NKPA III Corps — who were conducting harassing

Above: A sniper and spotter of 1st Marine Division in search of prey.
Right: Marines of 1 Battery, 11th Marines, in action with a 105-mm howitzer in support of 7th Marines in the Yanggu region of North Korea, June 1951.

operations between Andong and Usong. By the beginning of February 1951, the Marines' "Rice Paddy Patrols" had eliminated about 60 percent of the enemy and forced the remainder to pull out of the area.

At the end of February, General Ridgway began a series of limited counterstrokes, aimed at inflicting maximum damage on the enemy at minimum cost to the UN forces. The Marines were transferred to central Korea, where they came under the operational control of IX Corps in Operation Killer, which for public relations was subsequently renamed Operation Ripper. Two weeks later the shattered shell of Seoul was retaken by UN forces, the fourth time it had changed hands in less than nine months. In the central zone, the Marines secured Hungchon and destroyed the important Communist supply and communications center at Chunchon. On 4 April they were among the first UN forces to recross the 38th parallel, probing at a major Communist assembly and supply area between Chorwon, Kumhwa and Pyongyang, known as the "Iron Triangle".

As heavy fighting continued on the battlefield, a different kind of battle reached its climax. On 11 April President Truman replaced General MacArthur. The latter had been chafing at the restrictions placed on him by the UN and Truman; he had not, for example, been allowed to bomb bridges over the Yalu or fly reconnaissance missions over south China, for fear of precipitating full-scale Chinese involvement. The UN commander-in-chief was at loggerheads with the President over the conduct of the war. American B-29s and fighter-bombers had flattened North Korea's relatively primitive industrial infrastructure. Now MacArthur urged a naval blockade of China and a strategic bombing campaign against her industrial complexes in Manchuria. He argued that these operations should be combined with the employment of Chinese Nationalist forces from Formosa and the waging of a guerrilla war in south China. Truman was determined to limit the war to the Korean peninsula. Having abandoned any realistic hope of uniting North and South Korea, the United States and its UN allies were now reconciled to ending the war by negotiation. Relations completely broke down

Above: Clad in heavy winter uniforms, Corporal Richard J. Grinnic and Pfc Appleton, 1st Marine Division, undertake a scout-sniper mission in December 1951. Appleton is armed with a .30-caliber M1 Garand rifle with a sniper 'scope.

when MacArthur decided to pre-empt Truman with a public statement of his own, in which he declared that "there is no substitute for victory". Truman had no option but to remove his insubordinate commander-in-chief, replacing him with General Ridgway. Lieutenant-General James A. Van Fleet was despatched to Korea to command UN ground forces.

Van Fleet arrived on 14 April, a week before the Chinese launched their expected spring offensive. By the end of the month it had been contained, and the Communists had suffered 70,000 casualties. The second phase of the offensive opened on 15 May, the main blow falling on the eastern side of the Korean peninsula. After heavy fighting, the attacks petered out on 20 May. Eighth Army immediately counter-attacked, clearing South Korea of practically all the Communist forces by the end of the month. Some of the hardest fighting in Operation Piledriver was experienced by 1st Marine Division and 5th and 3rd ROK Divisions as they advanced on the "Punchbowl", a volcanic crater lying about 25 miles north of Inje in east-central Korea and along the edge of the "Kansas Line", Van Fleet's main defensive line.

As the fighting congealed into a slogging match around the 38th parallel, the Soviet representative at the UN, Jacob Malik, proposed a cease-fire. It was clear that the request had come from the People's Republic of China. Truce talks began on 10 July at Kaejing, just inside the Communist lines. The talks gave the Communists a much-needed breathing space, but achieved little else, and they were broken off in August.

Hostilities continued throughout the negotiations. At this point the front extended from the Imjin River to Chorwon, parallel to the base of the "Iron Triangle", swinging south-east along the southern edge of the "Punchbowl", then north again to come to rest on the Sea of Japan above Kansong.

Some of the fiercest fighting took place in the "Punchbowl", and it was here that the 1st Marine Division opened a new chapter in military history. At the beginning of the war, 1st Provisional Brigade had used Sikorsky HO3S-1 light helicopters in combat for the first time in the Pusan perimeter, flying reconnaissance, casualty evacuation and rescue missions. Unit commanders had used

Above: Improvizing front skids for wheels, Captain Grady Holmes, 1st Marine Air Wing, flies his Sikorsky HRS-1 helicopter on a shuttle run between forward and rear areas. In Korea helicopters pioneered airborne assault, rapid battlefield resupply and speedy relocation of mortars and light artillery, often within only a few yards of hostile troops.

them at Inchon to direct fighting over a front of several hundred yards. At the end of August 1951, a Marine medium helicopter squadron, HMR-161 commanded by Lieutenant-Colonel George W. Herring, arrived in Korea equipped with 15 Sikorsky HRS-1s. The HRS-1's sea-level payload of 1,420lb enabled it to carry up to six combat-equipped Marines or up to five casualties on litters. In the hand-to-hand combat over the razor-sharp ridges of the "Punchbowl", the helicopters offered an invaluable tactical advantage, with their ability to ferry quantities of supplies and troops in minutes over terrain which would take several hours to cover on foot or by motor transport.

On 13 September, Herring's HRS-1s were used in the first helicopter re-supply mission, delivering over seven tons of supplies (mostly ammunition) to 2nd Battalion, 1st Marines on Hill 673. Soon the helicopters were flying company-sized units to the combat-zone. A squadron of HRS-1s could carry a company 10–15 miles over terrain which would otherwise take a day to cover. In October the Marines began to move whole battalions by helicopter. The era of air mobility had begun.

The winter of 1951–52 passed with action on both sides confined mainly to raids and patrol skirmishes. In March 1952, after seven months hard campaigning in the "Punchbowl", the Marines were moved west to form part of I Corps, on Eighth Army's left flank, occupying a 35-mile front blocking the approach to Seoul.

From the cessation of major offensive action on 12 November 1951, the Main Line of Resistance (MLR) of the UN forces in Korea was a trench line which in theory ran all the way across the peninsula (the British Commonwealth Division broke the line with an "island" defensive system). The Marines' field fortifications were less impressive than those of the Chinese, behind whose deep-dug MLR complexes stretched echeloned belts of fieldworks some 10 miles deep and as impressive as anything on the Hindenburg Line in 1917. Comparisons with World War One were inevitable in these conditions, although this time the Marines were defending rather than attacking.

Just as in the trenches of 1914–18, there were long periods of inactivity — while both sides considered the results of the stop-go truce talks at Panmunjon — interrupted by bursts of savage fighting. Small, vicious battles were fought for the exchange of a few hundred yards of shell-churned earth or a hill-top which offered a few precious metres of elevation. A typical action was the bitter eight-day battle for the Chinese occupied Hill 122, a feature in no-man's-land dubbed "Bunker Hill" by the Marines and providing a vantage point from which the enemy was directing fire on the Marines' MLR. In a series of bloody actions between 11 and 17 August 1952, units of 1st Marine Division seized the hill at a cost of 48 dead and 313 wounded. Chinese casualties were estimated at about 3,000.

In March 1953, Communist pressure against the MLR began to build up to the level of May 1951, before the start of cease-fire talks. Hand-to-hand fighting occurred at Bunker Hill and other Marine outposts as the enemy probed at I Corps' line. Heavy fighting continued until July. The deadlock was broken by Truman's successor, President Eisenhower, who dropped thinly veiled hints that the United States was prepared to use nuclear weapons if peace talks were not resumed. On 20 July the talks got underway again at Panumnjon and an armistice was signed on the 27th.

Thereafter the Marines were relieved of their combat duties but remained in Korea until April 1955, manning defensive positions

Above: Medical corpsmen rush a wounded Marine from a Bell H-13 Sioux helicopter on a casualty evacuation (CASEVAC) mission.
Left: An uneasy moment for a North Korean farmer as he is interrogated by men of 1st Marine Division in April 1951.

Above: A battery of 4.5-in rockets unleashes a salvo at positions occupied by the Chinese Communist Forces (CCF). Rockets were an important supplement to the Marines 105-mm howitzers in the "war of the outposts", which was punctuated by ferocious artillery duels.

Right: An HRS-1 helicopter on a resupply mission in Korea.

along the Demilitarized Zone which separated North and South Korea.

Total US casualties in the war numbered approximately 136,000 men killed, missing or wounded. Marine casualties were listed as 28,011: 3,845 killed, 422 missing and 23,744 wounded. One-third of the Marine dead were victims of the static "outpost warfare" which had been such a feature of the war. Marine aircraft had provided vital close support, particularly during the withdrawal from Chosin. Between August 1950 and 27 July 1953, units of the 1st Marine Aircraft Wing flew more than 118,000 sorties, of which over 39,500 were close air support missions. During the same period, Marine helicopters had evacuated nearly 10,000 personnel.

The Korean war had given the Marines a taste of trench warfare which would have been familiar to the veterans of Belleau Wood. The landings at Inchon had drawn on the experience gained in the "island-hopping" campaign in the Pacific, while the "Rice Paddy Patrols" and the development of embryonic air mobile tactics anticipated a future conflict — in Vietnam.

VIETNAM

During World War Two, Allied support for resistance to the Japanese in South-East Asia ignited the twin forces of nationalism and Communism, the political effects of which are still working themselves out in the world of the 1980s. In Vietnam these forces were harnessed by the leader of the Viet Minh, the Communist Ho Chi Minh, who on 2 September 1945, from the northern city of Hanoi, issued a Declaration of Independence. A memorable photograph taken on the day shows the Vietnamese Communists giving their clenched-fist salute as the new red flag with the yellow star unfurled above the citadel of the former French colony of Indochina. By the side of these new leaders that day, also saluting the flag, were American officers in uniform.

The Americans would return, in less friendly circumstances. But first came the French, attempting to restore their colonial authority over Vietnam. Nine years later they were forced to admit defeat when 15,000 men, under General de Castries, were isolated and then overwhelmed at Dien Bien Phu by General Giap's Vietnamese People's Army.

At the Geneva Conference (26 April–21 June 1954) the French agreed to the division of Vietnam at the 17th parallel. Neighbouring Laos and Cambodia were recognized as independent, neutral states. The United States accepted the provisions of the Geneva Agreement but reserved the right to take action should they be broken. In North Vietnam a Communist government was established under the leadership of Ho Chi Minh. In the South a corrupt, authoritarian regime, under Ngo Dinh Diem, was set up with US backing.

Between 1954 and 1962 the US Marines provided a small advisory group to work with the South Vietnamese Marine Corps. During this period the situation in the South steadily deteriorated. Lacking popular support, Diem was unable to resolve South Vietnam's internal problems or counter the growing threat posed by Communist guerrilla forces, which in December 1960 formed the National Front for the Liberation of South Vietnam (NLF), called the Viet Cong (VC) by its opponents. By 1960 there were about 6,000 Viet Cong guerrillas operating in South Vietnam. However, the South Vietnamese Army (ARVN), advised by the Americans, continued to train for conventional rather than counter-insurgency operations.

On 19 October 1961, Diem declared a state of emergency. In November, following a fact-finding mission to South Vietnam by General Maxwell D. Taylor, President Kennedy decided to increase military aid without committing US troops to combat.

Early in 1962, Marine Medium Helicopter Squadron 362 of of 1st Marine Aircraft Wing was ordered to Vietnam to provide support for the AVRN. By now there were 4,000 US military personnel in South Vietnam, of whom 600 were Marines.

The United States had long since lost faith in Diem, and gave tacit support to his assassination on 1 November 1963 by a group of South Vietnamese generals. Three weeks later President Kennedy was also assassinated, in Dallas, Texas, and was succeeded by Lyndon Baines Johnson. In South Vietnam, nine changes of government took place in the next two years as coup followed coup. With the South on the verge of collapse, the decision was taken to initiate active US military intervention. This

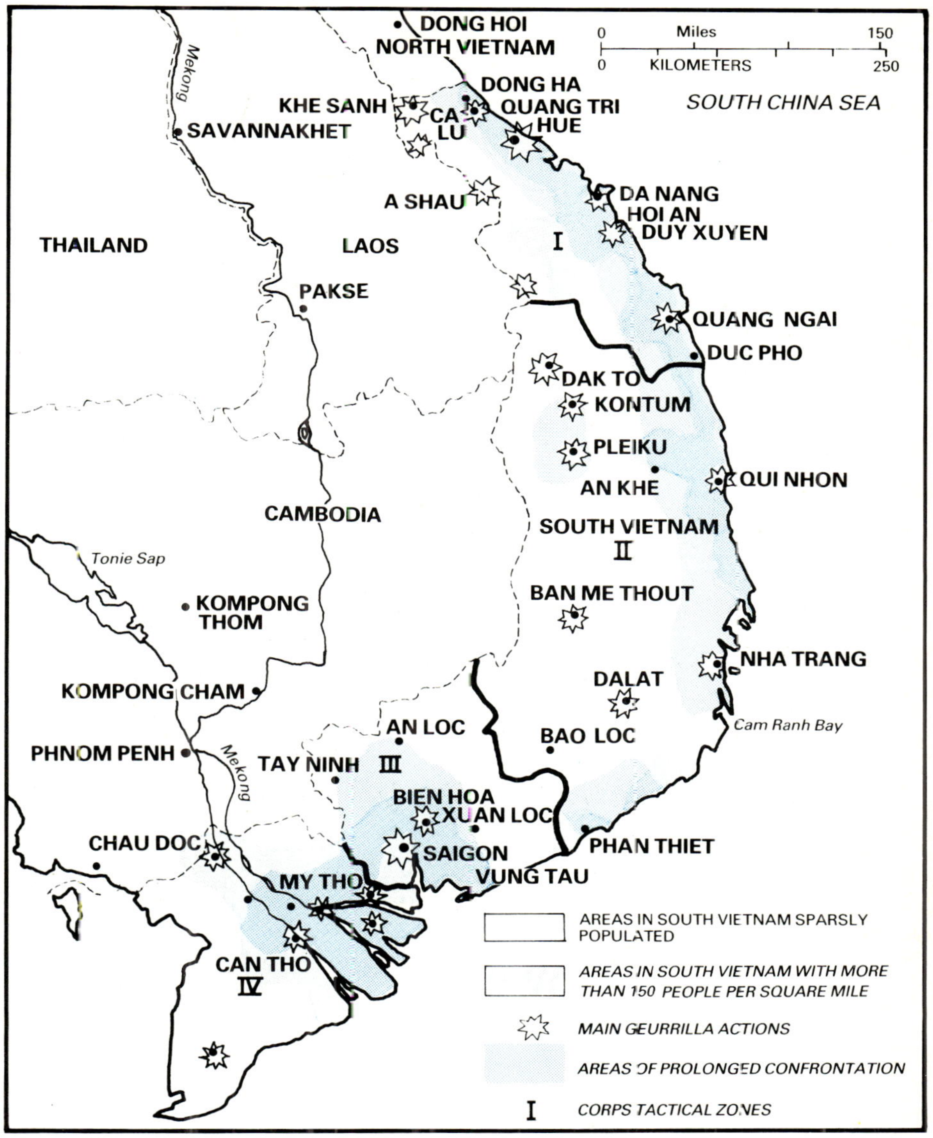

Above: The military regions of South Vietnam. Climate and topography had an important influence on the fighting in Vietnam. The Marines operated in I Corps Tactical Zone, where the monsoon occurred in the winter months. During the monsoon, visibility is poor, cross-country movement difficult and flying conditions bad. Health hazards — particularly for non-Asian troops — increased during the monsoon and quartermaster and storage problems multiplied. The North Vietnamese Army and the Viet Cong tended to reduce their levels of activity during the monsoon.

Above: Getting in. The 9th Marine Expeditionary Brigade comes ashore at Da Nang, South Vietnam, in March 1965. The Vietnam conflict was to be the longest, and in some respects the biggest, war in Marine Corps history. By 1968 III Marine Amphibious Force (III MAF) had a total of 85,755 men in Vietnam, more than were ashore at Okinawa.

resulted in the "Americanization" of the war, which was accomplished in successive stages between August 1964 and June 1965.

On 2 August 1964 the US destroyer *Maddox* was attacked in the Gulf of Tonkin by North Vietnamese torpedo boats. Two days later the destroyer *C. Turner Joy* also came under attack. On 5 August, aircraft from carriers of the US 7th Fleet retaliated by attacking military targets in North Vietnam. On 7 August, the US Congress approved the Gulf of Tonkin resolution, allowing the President to "take all necessary measures to repel any armored attack against the forces of the United States and to prevent further aggression". President Johnson now had a blanket authorization to expand the US military commitment in South-East Asia. By 1 January 1965, US military strength in South Vietnam had risen to 25,000 men.

In February, Viet Cong attacks on US military bases and personnel led to Operation Flaming Dart, in which carrier aircraft from 7th Fleet's Task Force 77 struck at military installations around the Demilitarized Zone (DMZ), the five-mile-wide buffer zone straddling the 17th parallel.

In March the role of US troops in South Vietnam was switched from "advisory" to "combat" and a sustained air offensive — Operation Rolling Thunder — was launched against North Vietnam.

Task Force 77 was not the only active force in the South China Sea. Since January, US Marines of the 9th Marine Expeditionary Brigade (9th MEB), in Task Force 76 shipping, had been steaming off South Vietnam, ready to land. At the end of February, President Johnson ordered the two-battalion brigade ashore to protect the air base at Da Nang, an operation which was already the subject of the Corps Schools' Amphibious Warfare Study No. 16.

At 0918 hrs on 8 March, the first Marines of Battalion Landing Team 3/9 (BLT 3/9) came ashore, to be greeted with garlands of flowers. In the afternoon the Brigade's second battalion, BLT 1/3, was flown in from Okinawa. By 12 March, 9th MEB was in position at Da Nang.

They were followed on 10 April by another Marine battalion, BLT 2/3. Two companies were immediately heli-lifted to defend the airfield at Phu Bai, 42 miles north-west of Da Nang and six miles from the ancient city of Hué. On 11 April, the first Marine fixed-wing

Left: McDonnell Douglas A-4E Skyhawk aircraft of Marine Attack Squadron 311 (VMA-311), armed with 750-lb bombs. In spite of its small size, the A-4 packed a considerable punch and, like all classic combat aircraft, was capable of absorbing continual updating and modifications.

Below left: Troops of 2nd Battalion, 4th Marines, move across a dike during a patrol of the Ly Tin area, north-west of the air base at Chu Lai, in 1966. At the beginning of their involvement in Vietnam, the Marines favored a strategy which gave priority to the establishment of secure coastal positions such as Da Nang from which forces could move out gradually in "clear and hold" operations. In contrast, the Army strategy increasingly favored the aggressive employment of troops in "search and destroy" missions against the enemy.

squadron, Marine Fighter Attack Squadron 531 (VMFA-531), flying McDonnell Douglas F-4B Phantoms, arrived at Da Nang. On 14 April, BLT 3/4 arrived to replace BLT 2/3's companies at Phu Bai. By the end of April, 8,878 Marines were in place in Vietnam.

The two Marine enclaves were situated in the northernmost of South Vietnam's three military regions, designated I Corps Tactical Zone and covering some 10,000 square miles of territory separated from the rest of the South by the rugged peaks of the Annamite Chain. Military control of this predominantly rural region was nominally by ARVN units commanded by General Nguyen Chanh Thi, but in fact their writ ran no further than the cities of Da Nang and Hoi Ann. In the countryside, a number of outpost district headquarters — garrisoned by company-sized ARVN units — floated like small islands in a sea dominated by the Viet Cong.

At the beginning of May, 9th MEB was redesignated III Marine Amphibious Force (III MAF), principally because the US Joint Chiefs of Staff were concerned that the term "Expeditionary" evoked memories of the colonial days of the French Expeditionary Corps. At the same time, it was reinforced by the landing of three more battalions at Chu Lai, 55 miles south-west of Da Nang and the site of a projected airfield.

Within 48 hours, Marine engineers and Seabees were working on the site, which became operational at the beginning of June. The airfield at Chu Lai was a unique Marine development, designated a Short Airfield for Tactical Support (SATS). This was nothing less than an "aircraft carrier on land", consisting of 4,000ft of aluminum matting runway, complete with catapult and arrester wires (the catapult was not installed at Chu Lai until 1967, jet-assisted take-off being used in the interim).

On 6 May, Da Nang-based Marine VMFA-531 Phantoms flew their first mission over North Vietnam and by mid-1965 there were four Marine Aircraft Groups — MAG-11, 12, 16 and 36 — stationed in Vietnam, two of them helicopter and two fixed-wing.

On to the Offensive

The American military build-up in Vietnam was inexorably gathering pace, and on 28 July President Johnson announced that the US forces in the region were to be increased from 75,000 to 125,000. The increasing demands of the US war machine led directly to an expansion of the Marine Corps, which late in 1965 received authorization for an additional 55,000 men. The Corps set a target of 278,184 men by mid-1967, and in the course of 1966 recruited 80,000 volunteers plus about 19,000 draftees.

In the summer of 1965 the Marines' roles in Vietnam underwent a radical change. Initially they had been required to adopt an "enclave strategy", standing on the defensive and patrolling within a radius of 50 miles around the enclaves to break up the concentration of hostile forces. In June 1965, President Johnson abandoned this strategy, secretly authorizing General William Westmoreland, Commander United States Military Assistance Command, Vietnam (Com USMACV), to engage in "counter-insurgency" combat operations.

At the beginning of August 1965 there were four Marine regiments — 3rd, 4th, 7th and 9th Marines — in Vietnam. Intelligence reports indicated that the 1st Viet Cong Regiment was concentrating on the Van Tuong peninsula, 15 miles south of Chu Lai. On 6 August, III MAF's commander, Lieutenant-General Lewis W. Walt, was authorized to take offensive action in Operation Satellite. However, a clerk's typing error transformed "Satellite" into "Starlite", and that was the way it stayed when the Marines jumped off on 18 August. In the first regimental-sized Marine action of the war, 7th Marines encirled and then severely mauled 1st VC Regiment, killing 614 of the enemy at a cost of 45 Marines dead.

The Viet Cong hit back on the night of 27-28 October when Communist sappers, covered by mortars, blew up 24 helicopters and damaged 23 more at the Marble Mountain Marine helicopter base, south of Da Nang. At Chu Lai, a VC raiding party destroyed two Marine Douglas A-4 Skyhawk fighter-bombers and damaged six more.

January 1966 saw the arrival in Vietnam of the 1st Marine Division. At the same time the Marines' Combined Action Program (CAP), initiated in the Phu Bai area, was extended to the Da Nang sector. The basic tactical unit of the CAP was the Combined Action Company, comprising a squad of Marine volunteers serving in each of the five platoons of local Popular Forces (PF) companies. The Marines were enthusiastic advocates of "winning the hearts and minds"

of the Vietnamese with a vigorous "pacification" program conducted at grass roots level, an approach which the Army treated with some scepticism. In spite of III MAF's considerable efforts in this field, the program was fatally undermined by the inability of the Popular Forces to provide security for an area once it had been cleared of Viet Cong.

The ultimate failure of the "pacification" program can be gauged from the grim fate of the village of My Lai, a hamlet near the provincial capital of Quang Ngai. In 1965 the Marines turned My Lai into a "model protected hamlet". On 16 March 1968 it was the scene of a massacre in which in an inexperienced, ill-disciplined company of 1st Battalion, 20th Infantry, US Americal Division, commanded by Lieutenant William Calley, ran amok, killing 347 Vietnamese civilians.

In January 1966, the four battalions of Marine Task Force Delta combined with the Army's 1st Air Cavalry Division in Operation Double Eagle, launched against the 325A North Vietnamese Army (NVA) Division, which was concentrating on the boundary between I and II Corps Tactical Zones. The operation began with 3rd Battalion, 1st Marines, landing near Thach Tru in the largest amphibious operation of the war up to that time, deploying three attack transports, an attack cargo ship, three LSTs, three LSDs, an LPH, a cruiser, a destroyer and two auxiliaries. On the second day of the operation, 2nd Battalion, 3rd Marines, was heli-lifted to a site five miles west of the beaches. On D+4, 2nd Battalion, 9th Marines, moved out of the Quang Ngai airstrip into the mountains to the north-west of the beaches.

Top: A VC guerilla is interrogated by men of 3rd Marine Division in 1966.
Above: Lance Corporal Miller Brown, 3rd Marine Division, fires his flamethrower into a VC position during Operation Cherokee (May 1966), 15 miles north of Hué.

Opposite, top: A flame-throwing tank in action, with Corporal Wayne Whiting surveying his handiwork. *Opposite, bottom:* Lance Corporal Jerry Fola, a rifleman in 3 Platoon, E Company, 2nd Battalion, 3rd Marines, at the alert during operation No Name Number 97, December 1967. He is armed with an M16A1 rifle, which can be distinguished from the original M16 by the prominent bolt-closure device above the pistol grip. This was added after combat experience in Vietnam revealed the difficulty of closing the bolt after debris had entered the weapon.

The enemy proved elusive, and in mid-February the Marines moved into the Que San Valley to engage 36th NVA Regiment and the reconstituted 1st VC Regiment in Operation Utah.

In March, the Marines launched two combined operations with the AVRN: Utah/Lien Ket 26, north-west of Quang Ngai; and Texas/Lien Ket 28, to relieve the besieged ARVN outpost at An Hoa. Over 1,000 enemy troops were killed in these operations, but they were overshadowed by the outbreak of large-scale civil unrest in South Vietnam's principal cities as Buddhists mounted a challenge to the military government. On 9 April, Marines at Da Nang had to intervene to prevent rebel ARVN units joining the dissidents. General Walt played a key role in restoring order with the minimum of bloodshed, but a major casualty in the disturbances was the "pacification" program, as Viet Cong infiltrated back into previously "sanitized" hamlets.

Below: A member of a Marine mine-sweep team carefully probes for hidden VC mines along the strategically vital Route 9 highway to the combat base at Khe Sanh, April 1968. Mines and booby traps accounted for some 11 percent of US combat deaths and 15 percent of wounds in Vietnam. A favorite spot for a mine was around a fallen tree or log lying across a track.

In March 1966, two-thirds of 1st Marine Division, commanded by Major-General Lewis J. Fields, was assigned to Quang Tin and Quang Ngai provinces. The 3rd Marine Division, under Major-General Wood B. Kyle, assumed responsibility for the Da Nang-Phu Bai region. By July it was clear that the North Vietnamese 324B Division had crossed the DMZ into northern Quang Tri province. Thus, in the summer of 1966, III MAF's ground war separated into two distinct campaigns: 3rd Marine Division's tactical war in the sparsely-populated northern I Corps area; and 1st Marine Division's "pacification" operations in the more heavily populated areas south from Da Nang. These rich ricelands were vital to the Viet Cong effort: they supplied their food needs; the adjacent coastline facilitated resupply from North Vietnam; and the dense population made it easier for their cadres to "swim like fish" in the sea of local inhabitants.

On 15 July, seven US Marine and five ARVN battalions launched Operation Hastings against the 324B NVA Division in Quang Tri province. Bombing support missions were flown by ground-directed Boeing B-52 Stratofortress aircraft of Strategic Air Command. During some of the heaviest large-scale fighting in the war to date, Marines occupied the "Rockpile", a 700ft vantage point from which they could observe west-central Quang Tri province. When the operation ended on 3 August, three Marine battalions remained in the area to seek out and engage NVA regulars in Operation Prairie, which continued until 31 May 1967.

By the autumn of 1966, the Marine presence in Vietnam had increased to 60,000 men. Its zone of responsibility now covered 1,800 square miles containing nearly one million people. It had participated in over 150 operations of battalion or regimental size, killing 7,500 of the enemy; another 4,000 enemy deaths had resulted from 200,000 Marine patrols, ambushes and other small-unit actions. Marine losses in this period were 1,700 killed and 9,000 wounded; over 80 percent of the wounded returned to duty.

Particularly successful had been the Marines' refinement of reconnaissance techniques, formalized under the code-name "Sting Ray", in which seven-man teams were used to observe the enemy and then call in fire missions from artillery, fighter-bombers and helicopter gunships. Two artillery pieces were assigned in direct support of each active Sting Ray team. By the time of the US withdrawal from Vietnam, Marine Reconnaissance teams were operating at will behind enemy lines. An outstanding pioneer of these techniques was Major (later Lieutenant-General) Bernard Trainor, commander of 1st Reconnaissance Squadron of the 1st Marine Division.

Operation Prairie I was brought to an end by the monsoon season at the end of January

1967. In 182 days of hard fighting around Con Thien and Gio Linh, the Marines killed 1,397 of the enemy at a cost of 225 killed and 1,159 wounded. Prairie II began in the same area in February, although on a much-reduced scale. The NVA had withdrawn across the DMZ and were infiltrating South Vietnam through the Ho Chi Minh Trail, a network of paths and trails through the jungle and mountainous terrain of Laos' eastern border, later extended into Cambodia. The Marines' role changed from "search and destroy" to "clear and secure", and the emphasis was placed on supporting the "pacification" efforts of 1st AVRN Division.

Spring 1967 saw the so-called First Battle of Khe Sanh, fought near the US firebase located in the north-west of I Corps Tactical Zone, six miles from the Laotian border and 14 miles south of the DMZ. This encounter battle began when a platoon-sized patrol from 1st Battalion, 9th Marines, made contact with an enemy force occupying the high ground to the north-west of the firebase. The

Above: A VC rocket attack lights up the sky over the Da Nang air base on 30 January 1968, at the start of the Tet offensive.

enemy did not return fire, but when the Marine patrol was joined by a second platoon, a furious firefight erupted in which 13 Marines were killed. On the following day it became clear that the enemy was well dug-in, to at least battalion strength, on three peaks, Hills 861, 881 South (881S) and 881 North (881N), apparently as a preliminary to attacking Khe Sanh.

Two Marine battalions were flown in, and the next 18 days saw a fierce battle in which enemy positions were pounded by artillery and air strikes delivered by 1st Marine Aircraft Wing. The Marines fought their way up the slopes and occupied the three peaks, establishing outposts protecting the approaches to Khe Sanh.

Following the First Battle of Khe Sanh, the Marines moved on to the offensive. In the south of I Corps Tactical Zone, 1st Marine Division launched Operations Union I and II, claiming 2,250 dead. There was also heavy fighting around the Marines' forward base at Con Thien, which came under sustained 122-mm and 140-mm rocket fire as the enemy concentrated for an assault.

War correspondents, ever quick to draw an historical parallel, began describing the fighting at Con Thien as the preliminary to a "second Dien Bien Phu". But to Generals Walt (who was succeeded by Lieutenant-General Robert E. Cushman on 1 June) and Westmoreland, the security of Con Thien was underwritten by American firepower. During the fighting at Con Thien, this was co-ordinated by the commander of US 7th Air Force, General William H. Momyer. From his forward headquarters he orchestrated the full range of fire support — tactical aircraft, naval gunfire and ground-directed B-52s, in close co-ordination with artillery and other ground fire — to break up enemy concentrations as they formed. By mid-autumn the "siege" of Con Thien was over.

The Tet Offensive

In the autumn of 1967, a return to a semblance of constitutional government in South Vietnam was made with the election of General Nguyen Van Thieu as President, with Air Vice Marshal Nguyen Cao Ky as his deputy. US military strength in the South had now risen to 500,000.

In three years of "search and destroy" operations undertaken by US forces, the Viet Cong had taken grievous punishment, frequently in the ratio 10:1. By the beginning of 1968 the Americans were fighting far from the South's main population centres, near the borders of Laos and Cambodia and on the DMZ. General Westmoreland, confident that

Left: A Marine Corps M50 "Ontos" — designed as a tank destroyer, mounting six 106-mm recoilless rifles, but widely used for fire support in Vietnam.

Below: Marines of A Company, 1st Battalion, 1st Marines, fire from the window of a house during the bitter fighting in Hué.

his victory was in sight, was planning to reduce the number of US troops in South Vietnam.

His optimism was suddenly reduced to ashes. On 22 January, the Marine combat base at Khe Sanh came under a siege which was to last 77 days. At the end of the month the Viet Cong and the NVA launched the "Tet" (Buddhist New Year) Offensive with simultaneous attacks on cities and military installations throughout South Vietnam.

In Saigon, the US embassy came under attack. Savage fighting raged in the beautiful old city of Hué, which was entirely occupied by the enemy with the exception of the ARVN 3rd Division's headquarters inside the citadel and a compound housing US Army advisors.

In some of the most bitter fighting of the war, it took the ARVN division and three battalions of Marines three weeks to clear Hué's Old and New Cities. Most of the Marines at Hué were short-term enlisted men and, although they had been trained to fight in a number of environments, expected to engage the Viet Cong in the countryside, where they could always rely on superior firepower and mobility. In Hué, the fighting was at close quarters, in rubble-strewn, fire-swept streets where blanket shelling was of limited value and heavy overcast skies prevented air strikes. Nevertheless, during the battle Marine artillery fired, 18,091 rounds: high explosive, smoke, white phosphorus, illumination and CS gas. From the Seventh Fleet, three cruisers and five destroyers expended 5,191 rounds; although the weather limited their effectiveness, Marine aircraft flew 113 sorties.

Just as at Iwo Jima and Okinawa, the Marines had to prise a fanatically-

Above: An aerial view of Marine Fire Support Base (FSB) Cates. The FSB was a self-contained and self-defending artillery base from which infantry operations could be supported. A typical FSB deployed a battery of six M102 105-mm howitzers; an infantry company for local defense; four 81-mm mortars from an infantry battalion; and communications, medical and administrative personnel.

determined enemy from their positions one by one. The Marines had some support from M48 tanks, which acquitted themselves well in the street fighting, and from the "Ontos", thinly-armored tracked vehicles mounting six recoilless rifles. These directed a great weight of fire against defense positions at close range. The "Ontos" could penetrate areas denied to the tanks, darting in to deliver a salvo before retiring to cover.

By the end of the Tet Offensive, the Viet Cong and the NVA had lost 400,000 men killed. Only at Hué did they hold any objective for an appreciable length of time, and this may explain why the North Vietnamese failed to commit troops held in reserve to exploit successes. The NLF, in particular, had suffered a shattering military defeat and was never the same force again. But they gained a crucial psychological victory. "Pacification" and US control over South Vietnam's cities were shown to be an illusion. The Tet Offensive, although a defeat for the NVA, made the US withdrawal from Vietnam inevitable. General Westmoreland's request for another 200,000 men — half for immediate use and half as a strategic reserve — was refused, and on 10 April President Johnson announced that he would be replaced by his deputy, General Creighton Abrams.

Khe Sanh

Two days earlier, the siege at Khe Sanh had

come to an end. The Marine combat base at Khe Sanh had been established as one of a series of firebases known as the McNamara line (after the US Secretary of Defense, 1961–68) and stretching from the coast to the Laotian border. Its purpose was to cut troop and supply infiltration into South Vietnam from the North and to block North Vietnamese incursions from Laos along the strategically important Route 9. At the beginning of January 1968, US intelligence sources indicated that a North Vietnamese offensive in the northern provinces was imminent. Two NVA Divisions, 324B and 325C, were identified in the area.

On the base, the Marine commanding officer, Colonel David E. Lownds, had at his

Previous page, top: An M60 machine-gun team of 5th Marines in a firefight in 1968. Troops in Vietnam dubbed the M60 "The Pig", as it was trouble-prone and very heavy for a squad-support firearm.
Previous page, bottom: Marines train their guns on an enemy bunker, down which they have sent a captured VC to retrieve the bodies of his comrades.
Right: Men of 3rd Battalion, 7th Marines, cross a river. They are wearing standard olive-drab combat dress and flak jackets.

disposal 1st, 2nd and 3rd Battalions, 26th Marine Regiment and an artillery unit, 1st Battalion, 13th Marines. As the siege developed, reinforcements were provided by 1st Battalion, 9th Marines, and 37th Rangers, an ARVN unit inserted for purely political reasons. The 6,000 men ultimately deployed at Khe Sanh faced enemy forces estimated at somewhere between 15,000 and 20,000.

The battle opened in earnest on 22 January with a massive rocket, artillery and mortar bombardment by the NVA which blew up 1,340 tons of ammunition in the base's main dump. Throughout February the enemy continued to pound the base and the Marine positions on Hill 881(S), an outpost on the high ground to the north-west of Khe Sanh. Hill 881(S), held by 400 men of 26th Marines, had nearly 50 percent casualties but each morning raised the Stars and Stripes on an improvized flagpole and brought it down at night to the strains of the correct bugle call.

Helicopter Squadron HMM-262, flying Boeing Vertol CH-46 Sea Knight assault transports, lost half their helicopters in February and had to be replaced by HMM-364. Helicopter deliveries were made three times a day — at 0900, 1300 and 1700 hrs — by "super gaggles" of 12 CH-46s with a standard load of 3,000lb, escorted by two UH-1E gunships, controlled by two UH-1E "slicks" and preceded by up to 12 strike aircraft, usually Marine A-4s from Chu Lai.

At the same time concentrated fire support in Operation Niagara broke the back of the enemy's attacks on Khe Sanh. NVA columns were allowed to advance until their heads approached the US perimeter. Three 105-mm batteries within the base then opened up with fixed concentrations forming three sides of a box around the enemy column with the open end facing the base. A fourth battery fired a walking barrage, moving up and down the box. Simultaneously, two batteries of 175-mm guns from the firebases at the

Below: The concept of helicopter-borne air mobility (which originated in Korea) came of age in Vietnam. *Below left:* Pfc John R. Huddeston, "A" gunner on an M-60 tank with 1st Marine Division, sits on the parapet of a fighting hole and watches fellow Marines board a CH-53 Sea King helicopter to return to An Hoa after 90 days in the field.

Command of the high-ground surrounding the low-lying combat base — Hills 881(S), 861, 558 and 950 (the numbers indicate their height in metres) — was vital to the survival of the garrison at Khe Sanh.

The NVA's siege methods would have won praise from the great 17th-century siege-master, Vauban, as they dug trenches, zig-zag approaches and parallels. Sometimes the NVA would dig 1,000ft of trenches in a night. On the following day A-4 Skyhawks would fly in to saturate them with napalm. NVA infantry attacks were mainly confined to Marine outposts, where the fighting was particularly savage.

The keys to the defense of Khe Sanh were resupply and fire support. The land main supply route — Route 9 — was too open to ambush, with stretches rendered impassable by the north-east monsoon. By the middle of February the landing and take-off of fixed-wing transports from Khe Sanh's airstrip had become too risky an operation. A switch was made to delivering supplies at low-level by the Low Altitude Parachute Extraction System (LAPES) and the Ground Parachute Extraction System (GPES). In the former, the transport made a low-level approach and a parachute dragged the cargo across rollers out of the rear doors. In the latter, the aircraft came in low to snag an arresting line which in turn pulled the cargo out of the rear.

Marine helicopters flying resupply missions were also vulnerable. Marine

Right: A flame tank sprays napalm.

"Rockpile" and Camp Carroll fired fixed linear concentrations outside the inner box, while fighter-bombers and B-52s, under Combat Skypot radar control, delivered a rolling barrage of ordnance to obliterate the enemy's reserve. NVA troops spilling out of the open end of the box were mopped-up by Marine infantry.

In March, the B-52s began a series of massive close-in strikes, arriving in cells of three over the target area every 90 minutes to hit enemy positions within 1,000yds of Khe Sanh's outer defenses. In all, the B-52s completed 2,548 sorties, dropping 59,542 tons of bombs. Photo-reconnaissance showed that 274 North Vietnamese defensive positions had been completely destroyed and 67 damaged, while 17 weapon positions were annihilated and another eight put out of action. Bomber crews reported 1,382 secondary explosions and 108 secondary fires.

NVA attacks on Khe Sanh began to fall away as the US relief operation, Pegasus, was set in motion, spearheaded by 1st Air Cavalry Division. On 4 April, 1st Battalion, 9th Marines, launched an attack from Khe Sanh to the south-east, capturing the enemy-held Hill 471. Here they linked with 2nd Battalion, 12th Cavalry, on 6 April. The enemy began to withdraw, losing their last strongpoint — Hill 881N — to attack by 3rd Battalion, 26th Marines, on 14 April. Pegasus was declared officially over at 0800 hrs on 15 April.

It has never been clear whether General Giap's intention at Khe Sanh was to force the Marines into a re-run of Dien Bien Phu. If it was, he failed, losing over 10,000 men in the process. Between 19 January and 31 March, the Marines at Khe Sanh lost 199 men killed and 830 wounded. Another 92 Marines died in Operation Pegasus. Two months later, the firebase at Khe Sanh was dismantled and its ammunition and stores moved back to the now secure Route 9. The move seemed to epitomize the American dilemma in Vietnam — the expenditure of colossal resources to no readily-discernible end against an enemy prepared to accept losses which would have been intolerable in a Western army. Pegasus was followed in the Khe Sanh area by Scotland II, which continued until February 1969, by which time the Marines claimed to have killed 3,921 of the enemy.

General Abrams now curtailed "search and destroy" operations, employing mobile helicopter-borne forces to strike at enemy forces where they were located. Two-thirds of US troops in South Vietnam were now tied to the protection of the main population centres. The corrosive effects of the war, watched on television every night by millions of Americans, provoked powerful political reactions at home. The bombing of North Vietnam, in particular, stirred deep controversy, engendered by the image of a superpower belaboring a small country. Ironically, the bombing — which was frequently interrupted in response to North Vietnamese hints that they might negotiate — was seen as a sign of American weakness rather than strength. Moreover, its relatively slow build-up, hampered by close operational control from Washington, enabled the North Vietnamese to put their country on a war footing and establish a formidable defense system, which accounted for 938 US aircraft during Rolling Thunder. As with the strategic air offensives of World War Two, by making the conflict a "total war" for the North Vietnamese people, the bombing strengthened their resolve. In the United States it added fuel to the anti-war movement.

Disillusioned, President Johnson began to cast around for a way out of the war. On 31 March 1968 he announced that he did not intend to seek re-election as President. The bombing of North Vietnam was halted north of the 20th parallel. In May, peace talks began in Paris, chaining the United States to the negotiating table until the North Vietnamese were ready to launch a conventional invasion of the South. On 21 October, bombing of the North ceased completely, bringing an end to Rolling Thunder.

Below: A Marine Grumman A-6 Intruder all-weather low-altitude attack aircraft. Marine squadrons at Da Nang and Chu Lai flew the Intruder, whose principal missions were close support, all-weather and night attacks on enemy troop concentrations and night interdiction. A typical load was 30 500lb bombs.

250 CARTRIDGES

Right: A Marine checks in with his command post via field radio in Operation Urgent Fury, Grenada, October 1983. He is armed with an M-16A1, with improvized lanyard, and a small fragmentation grenade hanging from his webbing.

On 5 November Johnson was succeeded as President by Richard Milhous Nixon, who had pledged to seek an end to US involvement in the war. The "Americanization" of the conflict was put in reverse, and its "Vietnamization" began.

As the American commitment to South Vietnam began to wind down, III MAF launched its last major action of the war on 22 January 1969. Operation Dewey Canyon, in the Da Krong valley, Quang Tri province, was ironically perhaps the most successful regimental-sized action of the war, involving 9th Marine Regiment, two battalions of 1st ARVN Division and the entire resources of 1st Marine Aircraft Wing. Over 1,000 enemy were killed in the operation, which ended on 19 March, and great quantities of arms and ammunition captured.

Now the newsreel started running backwards. On 8 June 1969 President Nixon announced the first US troop withdrawal. In July, 9th Marines left Vietnam; the 3rd Marines followed in October and, by the end of October, 3rd Marine Division had also gone. In April 1971 the headquarters of III MAF, 1st Marine Division and 1st Marine Aircraft Wing left Da Nang. All that remained was 3rd Marine Amphibious Brigade. Ground and air operations ceased on 7 May and, by the end of July 1971, 3rd Marine Amphibious Brigade had sailed away. Only a handful of US Marines remained in Vietnam, as artillery observers, advisors and embassy guards. They had grandstand seats when Saigon fell to the NVA on 30 April 1975. Marine losses in Vietnam (1962–72) were 12,936 killed, 88,594 wounded and 26 captured. Aircraft losses (to October 1970) were 252 helicopters and 173 fixed-wing aircraft lost in combat; 172 helicopters and 81 fixed-wing aircraft lost in operations.

AFTER VIETNAM

The Vietnam war dealt a heavy blow to American prestige and self-confidence from which it took a long time to recover. For the Marine Corps it was doubly galling. They had maintained their proud boast that they had never lost a battle, but the United States had lost the war.

On 10 November 1975, the Marine Corps celebrated its 200th birthday. Amid the junketings, many gave pause for somber reflection about the future. The Corps had come a long way since the days of the "Original Eight", but it was now moving into an era of political and economic uncertainty in which many of its precepts could no longer be taken for granted.

Training of an exceptionally rigorous kind has always been one of the cornerstones of the Marine Corps' effectiveness. But changing attitudes to authority resulted in an accelerating desertion rate which by 1976 was running at 69.2 men per 1,000 enlistees, compared with 31.7 per 1,000 for the US Navy and 11.7 per 1,000 for the US Army. A disproportionately high black intake — 22 percent, compared with 12 percent of the overall population — reflected the economic problems afflicting the United States and the difficulty young blacks were experiencing in finding regular employment. In 1979 the new Commandant of the Corps, General Robert H. Barrow, briefly considered tackling these problems by cutting the authorized strength of the Corps from 195,000 men to 175,000.

Gone were the days when a Marine force posted offshore could restore order to a troublespot threatened by civil unrest or civil war. The United States was now obliged to move more cautiously in a world of political and strategic complexities beyond the scope of gunboat diplomacy.

The Caribbean

Nevertheless, when it came to securing the

Above: Marines gather round a Soviet AN-26 Curl transport aircraft of Cubana airlines seized at Pearls Airport during Operation Urgent Fury.

interests of the United States in a traditional sphere of influence, the Marine Corps could still be called upon to move as decisively as it did in the days before World War One.

In the Dominican Republic, the assassination of the pro-US dictator Rafael Trujillo on 30 May 1961 was followed by a prolonged period of political turmoil. In April 1965, a left-leaning faction in the Army overthrew the conservative Cahral regime. Fearful that the Dominican Republic might follow Cuba into the Soviet camp, President Johnson ordered a six-ship squadron of the US Navy's Caribbean Ready Group, with some 1,700 Marines, to sail from Puerto Rico, ostensibly to protect the lives of US nationals and evacuate them if necessary.

On 27 April, while fierce fighting raged in Santo Domingo between the leftist insurgents and a new military junta, Marines using helicopter and amphibious transports succeeded in evacuating 1,200 US nationals aboard US ships. The next day the Dominican junta requested military aid. The platoons of Marines which were sent ashore quickly came under fire from rebel snipers. In Washington the decision was taken to commit more Marines in order to prevent a Communist coup.

On the 29th, 1,500 Marines were landed, followed by elements of the US 82nd Airborne Division. Heavy fighting followed as they pushed into Santo Domingo. Eventually, the number of US troops in the Dominican Republic reached 32,000, including 14,000 Army personnel, 8,000 Marines, 9,000 Navy personnel and 1,000 airmen. On 6 May a peace force created by the Organisation of American States (OAS) came into being, at the request of the United States. Once the OAS units were in place in the Dominican Republic, the United States began to reduce its military presence. The first to go were the Marine battalions, and by early June over 10,000 troops had been withdrawn. The remainder left in the summer of 1966, following the election of a new President, Dr Joaquin Balaguer.

The Marines went into action again in the Caribbean in October 1983, on the tiny island of Grenada. In-fighting between competing factions in Grenada's Marxist government led to the seizure of power by an extreme left-winger, Bernard Coard, and the summary execution by firing squad of the Prime Minister, Maurice Bishop. Of particular concern to the United States was the 10,000ft runway under construction at Point Salines airfield on the western tip of the island. Funded with Russian money and built by Cuban combat construction workers, it threatened to provide the Soviet Union with a crucial toehold in the region from which it could "service" client states.

The Organisation of Central Caribbean States (OCCS), together with Jamaica and Barbados, called upon President Ronald Reagan to provide military intervention.

Reagan was not slow to comply. The chaotic situation on Grenada was seen as a direct threat to American security. By deploying a military force to eliminate the threat before it fully developed, Reagan hoped simultaneously to restore a pro-American government in a strategic location and also to demonstrate the United States' resolve to maintain the *status quo* in the Caribbean. A viable pretext for intervention was supplied by the presence on Grenada of several hundred US medical students.

The units selected for Operation Urgent Fury were US Army Rangers of the 75th Infantry (Ranger) Regiment; US Army paratroops of 82nd Airborne Division; and Marines of the 22nd Marine Amphibious Unit (MAU), who had been diverted from another troublespot, the Lebanon, to take part in the operation. Additionally, there were small contingents from special operations units: Navy SEALs, US Army Special Forces, USAF Combat Control Teams (CCTs), and the Delta anti-terrorist unit, which was standing by to act if any of the American students were taken hostage. A force of 400 police and military personnel was provided by Grenada's Caribbean neighbours, to act as a garrison after the island had been secured.

The operation began in the small hours of 25 October 1983 when SEAL units came ashore, tasked with the rescue of the island's Governor-General, Sir Paul Scoon, who had been placed under house arrest. At 0536 hrs, the Rangers made an airborne assault on the airfield at Point Salines. Simultaneously, 400 Marines from the assault ship *Guam* made a heli-borne landing at Pearls airport, on the eastern side of the island. Pearls airport was secured with little difficulty, but at Point Salines the Cubans, who were heavily armed, put up a stiff fight. However, the runway was cleared and operable by 1400 hrs, when two battalions of 82nd Airborne were flown in to relieve the Rangers and free them for other operations.

Meanwhile, *Guam* was moving round the island and at 1930 hrs landed 250 Marines, five tanks and 13 LVPT-7 amphibious assault vehicles at Grand Mal Bay, near St George's, the island's capital. Their objective was Fort Frederick and its Richmond Hill prison, on the high ground above St George's, where the SEALs and the Governor-General were now besieged.

At 0700 hrs on the 26th the Marines secured Fort Frederick, which had been abandoned by its defenders. Moving eastward from Point Salines, the Rangers and 82nd Airborne took Frequente and at 1600 hrs secured Grande Anse campus, where the majority of the students had taken refuge. The students were heli-lifted out on the same day. Throughout the rest of the 26th, men of 22nd MAU and 82nd Airborne swept the island for Cubans and troops of the Grenadan People's Revolutionary Army. A curfew was imposed during darkness.

By the third day of the invasion, all resistance had evaporated and the Cuban barracks at Edgmont were taken without a fight. Large quantities of light and heavy weapons were discovered, including AK-47 assault rifles, 120-mm mortars, machine-guns, anti-aircraft weapons and rocket launchers. Total US casualties in the operation were 18 dead and 113 injured. More than 1,100 Cubans and Grenadans had been taken prisoner, and Cuban casualties had been considerable. A week later all the US Rangers and Marines had left Grenada, and within six weeks there were only about 300 US military personnel left on the island.

Left: Marine Sea Stallion helicopters on the flight deck of the carrier *Nimitz* at the start of the ill-fated mission to rescue the US hostages in Teheran. The failure of the mission stirred memories of the "*Mayaguez* incident" of May 1975. On 12 May of that year a Cambodian gunboat had seized the American merchantman *Mayaguez*. An operation to recover the ship and rescue its crew (who had been imprisoned on the mainland at Koh Tang) resulted in the loss of four Marine helicopters and 18 men (11 of them Marines). Nevertheless, the crew and the ship were retrieved.

The Middle East

The politics of the Caribbean seem relatively uncomplicated when compared with the murderous complexities of the Middle East. On 16 January 1979, the Shah of Iran left the country in the face of relentless mass demonstrations. On 1 February, the Ayatollah Khomeini returned from exile to gather to himself the reins of power of the Islamic revolution.

On 4 November 1979, 400 militant Islamic students occupied the United States embassy in Teheran. Its Marine guards were among the 66 US citizens taken prisoner during the occupation. Three other US citizens were also held hostage in the Iranian Foreign Ministry.

Marine pilots were involved in an ill-fated American rescue bid of 14/15 April 1980 by Colonel Charles Beckwith's 1st Special Forces Operational Detachment (Delta).

The plan was for the Marines to fly eight Sikorsky RH-53 Sea Stallion helicopters from the carrier *Nimitz* to a secret rendezvous in the desert 200 miles south-east of Teheran with three troop-carrying Lockheed MC-130 Hercules transports accompanied by three fuel-carrying EC-130s. Delta's assault force of 118 men and their equipment were to be loaded on to the helicopters and, after refueling, flown on to a hide-site near Teheran, from which they would launch a complicated operation to free the hostages. Covered by the massive firepower of C-130 gunships, the Sea Stallions were to fly in to pick up the assault teams and hostages from the embassy compound or, if requested, from a nearby soccer stadium. The Sea Stallions would then ferry the hostages and assault teams to Manzariyeh airfield, 35 miles to the south, where a contingent of US Rangers would be defending a flight of Lockheed C-114 Starlifter transports waiting to airlift everyone out of Iran.

The plan broke down in chaos at the initial "Desert One" rendezvous south of Teheran. Only six helicopters arrived, an hour late and long after first light, having been delayed by severe dust storms which had forced one pilot back to the *Nimitz* with navigational and flight instrument failure; a second had come

Opposite: The USS *Fort Snelling*, a Thomaston class Landing Ship, Dock (LSD). *Fort Snelling* is capable of accommodating 21 LCM(6) landing craft or three LCUs (Landing Craft, Utility) and six LCMs, or 50 LVTP amphibious tractors, with another 30 LVTPs housed in the mezzanine and upper decks; troop capacity is 340.

Below: The AV-8 Harrier, which equips four Marine squadrons.

down with rotorblade trouble only 80 miles inside Iranian airspace.

After refueling and loading, a third helicopter suffered hydraulic failure and the mission was abandoned. The withdrawal plan was for everyone to offload and rejoin the transports. The five Sea Stallions would then fly back to *Nimitz*. As one of the Sea Stallions was changing position it collided with a troop-laden Hercules and burst into flames. In the ensuing conflagration, Redeye missiles exploded in all directions. Miraculously, all the troops inside the Hercules managed to disembark as the C-130 caught fire. Eight men crewing the C-130 and the Sea Stallion were killed. Boarding the remaining aircraft, Delta Force swiftly abandoned "Desert One", leaving five helicopters behind on the ground and the eight charred bodies of the men who had died in the mission.

The hostages were finally released in January 1981 after prolonged negotiations. The "military solution" had not only proved a humiliating failure for the United States — it had in the final analysis proved unnecessary. In the process, it dealt a terminal blow to the Presidency of Jimmy Carter.

The situation in the Persian Gulf was also part of an array of long-term strategic problems facing the United States. The violent anti-Americanism of the Khomeini regime, the prolonged hostage crisis, and the Soviet intervention in Afghanistan in November 1979, seemed to pose a direct

Below: Marine training has a well-earned reputation for toughness. The cream of the Corps volunteer for the Force Reconnaissance Units, whose task is to gather intelligence prior to a landing. "Recon" units are divided into Force Recon Company groups, which operate covertly in specialist four-man teams somewhat in the style of the British SAS, and Battalion Recon Company Groups. Battalion Recon is larger, each Division having a battalion of around 500 men. Divided into platoons and companies, its members are attached to specific battalions. Utilizing SCUBA diving techniques, helicopters and light vechicles, they obtain general intelligence regarding beaches to be used for landing by the parent battalion.

threat to Western oil supplied from the Gulf and the stability and security of the pro-Western states in the region. In his State of the Union message of 21 January 1981, President Carter made a clear commitment to the security of the Gulf, and indicated an American readiness to defend its interests there, if necessary by military force. Within a month a Marine Amphibious Task Force, including four vessels led by the amphibious assault ship *Okinawa* and carrying some 1,800 men, left the Pacific for the Indian Ocean for maneuvers with Carrier Task 70 on station in the Arabian Sea. Since then, similar Marine Task Forces have been rotated to the Arabian Sea to maintain a permanent presence there.

The deployment of the Marine Amphibious Task Force in the Arabian Sea was a significant step in the progress towards the creation of the Rapid Deployment Joint Task Force (RDJTF). The idea of a quick-reaction force, capable of worldwide deployment, had first been canvassed in the early 1960s by the US Secretary of Defense, Robert McNamara. Drawing from the lessons of the Cuban missile crisis of 1962, McNamara concluded that the advantage in sub-nuclear superpower confrontations would go to the side which could deploy a decisive local military superiority. Such a capability would also help prevent an escalation of such confrontations to the nuclear level.

The situation in the Middle East in the late 1970s gave the concept a new urgency. The Rapid Deployment Joint Task Force was established on 1 March 1980, with its HQ at McDill Air Base, Tampa, Florida; in October 1981 it became a separate task force with joint command and control of the forces designated to it; and on 1 January 1983 the RDJTF became a separate unified command, US Central Command (CENTCOM). The importance of the amphibious element of the RDJTF was indicated by the appointment as its first commander of Marine Lieutenant-General Paul X. Kelley. The Marine element of the RDJTF includes three Maritime Prepositioning Ship (MPS) task forces, each of which can carry sufficient supplies for a highly mechanized Marine Amphibious Brigade.

Left: One of the Marine Corps' 720 M-60A1 tanks, with wading exhaust.

Below: Marines armed with M-16s take cover behind sandbags in Beirut. The Corps had first fulfilled a peace-keeping role in the troubled Lebanon in July-September 1958, on that occasion suffering only two deaths, both accidental shootings by other Marines. Their second tour of duty, in 1984, was to prove more traumatic.

The Lebanon

Nowhere are the intractable problems of the Middle East more dramatically demonstrated than in the Lebanon, which since the summer of 1975 has been torn apart by a civil war between rival Christian and Islamic factions. The arrival of the Palestinian Liberation Organisation (PLO) in southern Lebanon injected a factor guaranteed to increase Lebanon's chronic instability. In mid-1976 the President of Lebanon appealed to Syria for help. Syrian intervention secured a temporary cessation of the fighting, but the warring factions failed to reach any agreement among themselves. While the Syrians remained the only guarantors of a semblance of internal security, they quickly became an army of occupation. A United Nations force — UNIFIL, based in southern Lebanon — was little more than a helpless bystander in the conflict.

Reacting to the Syrian presence, and determined to put an end once and for all to the PLO, the Israelis invaded Lebanon in June 1982, driving northward to Beirut. In August 1982, an agreement was negotiated by the US envoy Philip Habib, calling for an Israeli withdrawal from the Lebanese capital. In response to a request from the Lebanese government, an international peacekeeping

The shattered ruins of the compound in which 240 Marines died on 23 October 1983.

force, comprising French and Italian troops and 800 US Marines, arrived in Beirut at the end of August. By the beginning of September, all the Syrian and PLO fighting units had been evacuated from Beirut. On 13 September, the newly-elected President of the Lebanon, the Maronite Christian Bashir Gemayel, was assassinated with Syrian connivance.

On the following day the Israelis marched back into Beirut. They saw themselves as the only force for stability in an increasingly chaotic situation, but they immediately made a major miscalculation. On 16 September their troops stood idly by while Maronite Christian militiamen massacred Palestinian refugees in Beirut's Sabra and Shatila camps.

The US Marines had returned to their ships offshore on 10 September. Now President Reagan ordered them back into

Lebanon, and on 29 September they were detailed to guard Beirut International Airport. On 23 October a Muslim suicide bomber drove a truck packed with explosives into the Marine compound, where it blew up with tremendous force, killing 240 Marines. The Americans, naively perhaps, had seen their role as that of peacekeepers. But to the suicide bomber — a member of the extreme Iranian-backed Shi'ite Hizbollah faction — they were just another militia and representatives of "The Great Satan" — the United States. A fatal incomprehension of the tortuous coils of Lebanese politics, and insufficent vigilance, had led to one of the blackest days in the Marine Corps' history. The Marines left Beirut in the spring of 1984, having lost 260 men during their tour of duty.

Left: Marine sniper scouts survey an infinitely hostile terrain in Beirut.

Bottom: A Marine LVTP-7 tracked landing vehicle in the streets of Beirut. This amphibious assault vehicle can carry up to 25 fully-equipped Marines, propelling itself through the water with twin water-jets. The LVTP-7 is equipped with passive night vision equipment, smoke-generating capacity and automatic fire detection and suppression system. Armament is provided by a 12.7-mm M85 machine-gun. It has a 40mph road speed and 8.39mph water speed, with a road range of 300 miles.

Opposite, top: The USS *Inchon*, an Iwo Jima class LPH (Landing Platform, Helicopter). The seven LPH ships of the class, commissioned in 1961-79, were the world's first assault ships designed and constructed specifically to operate helicopters. The LPH can accommodate 11 CH-53 Sea Stallions or 20 Boeing-Vertol CH-46 Sea Knights. Each can carry a Marine battalion landing team, its artillery, vehicles and equipment plus a reinforced squadron of assault helicopters. Anti-aircraft defence is provided by Sea Sparrow SAMs and Phalanx CIWS. In the late 1980s each full-strength Marine battalion consists of a headquarters company, three rifle companies divided into 13-man squads, and a weapons company. The standard squad weapon is the improved M16A2 rifle supported by a 5.56-mm Squad Automatic Weapon (SAW) issued one per fire team. The weapons companies are acquiring a new heavy machine-gun platoon with eight firing teams each allotted a vehicle armed with a 0.5-in and a Mk19 40-mm heavy machine-gun.

MARINE TRAINING

At the Marine Corps Recruit Depots a gruelling training program and iron discipline transform new civilian trainees into closely-welded platoons of combat-ready fighting men. The recruits pass through three clearly-defined phases of training, the first of which can best be described as "shock treatment" — an intensive introduction to the fundamentals of military life. From their arrival in "boot camp", the recruits are put through a highly ritualized psychological assault course by their drill instructors (DIs), aimed at stripping away all vestiges of civilian attitudes and reducing the recruits to the same level. Complete denial of personal privacy, ruthless attention to the minutest details of dress and kit maintenance — all accompanied by the eyeball-to-eyeball verbal pyrotechnics of the DIs — combine to produce an automatic response to orders. "Problem" recruits are assigned to a special "motivation platoon", whose program includes the screening of inspirationl feature films like *The Sands of Iwo Jima*, starring John Wayne. Long hours of drilling on the parade ground (the "grinder") and punishing daily exercises rapidly increase discipline. The second phase concentrates on marksmanship and the maintenance of personal small arms. After three weeks on the range the recruits qualify as riflemen. Out of a possible 250 points (five points for each of the 50 rounds fired) a man needs 190 to qualify as a "marksman", 210 for "sharpshooter", and 220 for "expert". By now, competition between platoons encourages unit loyalty and pride. Non-qualifiers in the marksmanship course are often made to wear their shooting jackets backwards and march away from the range at the rear of their platoons, carrying their rifles upside down. After a week of mess and maintenance duty, the third phase begins. The DIs now foster maximum competition between the platoons for the prize of honor platoon on graduation day. This phase of training includes a day on the Confidence Course — an assault course designed to test the recruits' strength, agility and physical courage — and the issuing of metal identification tags ("dog tags"), which reinforces the recruits' growing sense of experience. The final days of recruit training are spent preparing for a series of written tests, which are followed by a last physical fitness test and the series drill competition, judged by senior NCOs. The last hurdle for every platoon is the "Final Field Inspection", by a team of officers and senior NCOs from the recruit training battalion. The subsequent graduation parade is conducted with great ceremony and watched by the recruits' families and friends. This is a marked change from the days in the 1950s when, on graduation, recruits simply packed their gear in a seabag and joined an operating unit.

Right: A Marine Corps Honor Guard. The Guard is wearing the Marines Dress Blue "B" uniforms. The "blood stripes" on the trouser seams of the corporal nearest the camera distinguish NCOs, WOs and officers from other ranks.

Below: A vision of the future — a Marine in an NBC (Nuclear-Biological-Chemical) warfare kit.

Looking to the Future

Throughout its history, the Marine Corps has resisted determined attacks on its independence. Today the Corps' powerful lobby in Congress ensures that it will retain its unique identity into the foreseeable future.

The active strength of the Corps is currently around 199,000, including 9,700 women, divided into three Marine Divisions, two Marine Security Force battalions and three active Air Wings. The Air Wings consist of over 40 squadrons of fixed and 30 squadrons of rotary wing aircraft. The 472 combat aircraft, together with the 84 armored and approximately 450 support helicopters, provide the Marines with a unique versatility. They are capable of undertaking a number of battlefield roles, including close support, transportation and combat air patrols. Mobility is provided by helicopter units capable of lifting men and equipment into and out of action. The Marine Air Wings also deploy early warning and reconnaissance aircraft. The concept of the all-arms force is embedded in the structure of the Marine division. In addition to its three rifle regiments, each division has an artillery regiment, a tank battalion, an amphibious battalion, a light armored assault battalion, reconnaissance battalion, air defence battalion, engineering battalion and other support and ancillary units. Although Marine Corps divisions are some 20 percent larger than their Army equivalents, financial constraints make their battalions smaller.

Despite the recent Soviet build-up of its amphibious forces and the technical innovations in equipping them, they remain small in comparison with the US Marines and the large number of specialized ships which support them, sufficient to lift one and one-third Marine Amphibious Forces (MAFs). A

MAF comprises three reinforced battalions, plus an air wing — a total of 45,000 Marines and 6,000 Navy personnel. Although the MAF is used as a standard measurement criterion for amphibious lift, a more realistic indicator is provided by the number of reinforced battalions which can be maintained in forward areas, primarily the Mediterranean, Western Pacific and Indian Oceans, supported by Maritime Prepositioning Ships. The US Navy is capable of keeping two reinforced battalions continuously afloat in the Western Pacific and one in the Mediterranean. In addition, a reinforced battalion is intermittently deployed in the Atlantic.

Among the 60 active and reserve ships are Amphibious Assault Ships (LHAs), Landing Platform Helicopter (LPHs), Command and Control Ships (LCCs), Amphibious Transport Docks (LPDs), Landing Ships, Dock (LSDs), Landing Ships, Tank (LSTs) and Amphibious Cargo Ships (LKAs).

The five ships of the "Tarawa" LHA class are the largest purpose-built assault ships in service, capable of accommodating a reinforced battalion, 19 Sea Stallion helicopters and a large number of fighting vehicles and palletized stores. Beneath the full-length flight decks are two half-length hangar decks connected by lifts. Under the aft elevator is a floodable dock capable of accommodating four LCU1610 landing craft. In addition there are extensive hospital facilities and storage for 10,000 gallons of vehicle fuel and 400,000 gallons of helicopter fuel.

In spite of this formidable display of military technology, the US Marine Corps remains faithful to the traditions which began in the fighting tops of tiny frigates during the American War of Independence. The future will see them on station, and on alert, throughout the world.

Index

Page numbers in *italics* refer to illustrations.